WHAT LIZARD IS THAT?

Introducing Australian Lizards

Steve Wilson and Gerry Swan

Published in Australia by Reed New Holland
an imprint of New Holland Publishers (Australia) Pty Ltd
Sydney • Auckland • London

Unit 1, 66 Gibbes Street Chatswood NSW 2067 Australia
5/39 Woodside Avenue, Northcote, Auckland 0627, New Zealand
The Chandlery, 50 Westminster Bridge Road, London SE1 7QY, United Kingdom

National Library of Australia Cataloguing-in-Publication data:

Wilson, Steve, 1954-
What lizard is that? / Steve Wilson.
9781877069581 (pbk.)
Lizards–Australia–Identification.

597.950994

Publisher: Fiona Schultz
Publishing manager: Lliane Clarke
Project editor: Diane Jardine
Designer: Barbara Cowan
Cover design: Tania Gomes
Production manager: Linda Bottari
Printer: Toppan Leefung Printing Limited (China)

Cover image: Painted Dragon (*Ctenophorus pictus*)
Title page image: Chameleon Gecko (*Carphodactylus laevis*)

10 9 8 7 6 5 4 3 2

If men were men as much as lizards are lizards
they'd be worth looking at.

D.H. Lawrence

contents

Introduction

The land of lizards

Lizards are so common and widespread in Australia that we accept them as a fundamental part of the landscape. Virtually every backyard harbours at least one, usually several, different kinds of skinks. They range from the alert little 'drop-tails' basking in rockeries and skittering off path edges to the large and ponderous blue-tongues that crush the snails in our garden beds and gape their flat tongues at passers-by. In some of the busiest Australian cities, fat, street-wise dragons laze beside ornamental ponds, while across the northern half of the continent house geckos scamper with gravity-defying ease across walls and ceilings to snatch the moths attracted to the lights at night.

That's just the human environment. On the peaks of the highest mountains there are lizards soaking up the meagre sunshine. Such places are relatively poor in species but often support abundant individuals that sometimes cluster together, pooling their acquired heat to create a larger, more thermally stable mass. Others dodge incoming waves on some of our remote rocky coastlines, adding marine invertebrates to their diet of insects.

Across the country, large goannas devour roadkill along outback highways and swagger across kilometres of terrain as they probe burrows and investigate nooks and crannies for prey. In the muted light of tropical rainforests, some skinks bathe in the puddles of sunshine while others lurk in the eternal dampness within rotting logs. As darkness sets in, leaf-tailed geckos emerge from rock cavities and hollows in the latticed fig trunks to stake out their vantage points on rock faces and tree buttresses. Head-downwards, with long limbs splayed spider-like, they are invisible against the moss, lichen and rough bark.

Some of the richest lizard habitats on Earth are Australia's deserts. Millions of square kilometres of dunes and sandy flats are scribed with the footprints and meandering squiggles of foraging lizards. Like a cycle of shift workers, the complex assemblage of species partitions their activity times to emerge at dawn, mid-morning, late afternoon and evening. Each clump of spinifex offers perfect lizard retreats, binding the sand to create sheltered burrowing sites, and enclosing a humid pocket of air protected by an interlocking matrix of tough, needle-sharp spines. Geckos and flap-footed lizards forage and feed within these clumps, and other lizards, including pygmy goannas and striped skinks, tunnel under them. This one vegetation type has been recognised as a major contributing factor to Australia's immense desert lizard diversity.

With nearly 650 named species in five families, Australia can truly be called the 'Land of Lizards'. Indeed, ours is the only continent where the largest mobile terrestrial predators capable of active pursuit are not mammals but lizards.

◄ A Leaf-tailed Gecko (*Phyllurus ossa*) rests artfully camouflaged against a lichen-encrusted boulder. Mt Ossa, Qld.

Secret lives

▲ The Thorny Devil (*Moloch horridus*) is a slow-moving, inoffensive inhabitant of sandy deserts. Simpson Desert, NT.

Some Australian lizards are so instantly recognised that they are hailed as fauna icons. Frilled Lizards grace highway signs, tourist brochures, advertising billboards and virtually every popular Australian wildlife picture book ever written. Likewise, the Thorny Devil is a big-ticket drawcard for tourists visiting the 'red centre'.

But the majority of lizards are not as famous. Scientists are still discovering and naming them, and for the most part their lives unfold in secret. We humans are privy to just a small glimpse of their extraordinary behaviour. Beyond passive sun-basking, our observations have much more to do with how lizards respond to us than how they interact with each other. We mainly see them in terms of 'flight or fright' – how they flee when pursued, break their tails if grasped, and sometimes stand their ground with impressive bluff if cornered.

Occasionally we chance upon spectacles such as combat – real or ritualised – between rival males. Dragons strike exaggerated poses, seeking dominance through posture and intimidation; monitors stand chest to chest, attempting to shove each other into submission; and geckos rear their bodies, utter warning calls and sometimes sever their opponents' tails. We may also be lucky to see courtship displays, or even come across a male grasping a female by the nape in the act of mating.

We rarely witness egg-laying though sometimes we encounter a bearded dragon digging its nest-burrow in a load of freshly delivered sand or tunnelling into the side of a bush track.

▲ The Southern Angle-headed Dragon (*Hypsilurus spinipes*) is confined to eastern Australia's subtropical rainforests. Mt Glorious, Qld.

And it is not uncommon to discover clutches of skink eggs, sometimes even communal caches of several dozen, when we move brick piles or disturb rockeries. Young are usually born and eggs hatch unseen by us. We note the event by the abrupt appearance of tiny lizards.

Lizards are abundant, thrive in our gardens and often enter our houses. They engage in fascinating antics, but the natural history of many species remains poorly known, and we are still a long way from establishing just how many different kinds actually live in Australia. Some undiscovered and unnamed species may even have vanished from under our noses without our having been aware they existed. Others are certainly under serious threat. For all these reasons, lizards are worthwhile subjects to study. Just keeping an eye on the extraordinary things they get up to in our own backyards offers rewarding insights into a whole different world unfolding around us.

About this book

▲ The Collared Delma (*Delma torquata*) is a small legless lizard endemic to south-eastern Queensland. Mt Crosby, Qld.

This book is intended as a broad introduction to Australia's diverse lizard fauna. It covers most groups of Australian lizards and includes general pointers on appearance, behaviour and ecology. Where identification features are of particular interest, then they are provided, but this book is not intended as a field guide. Such books are only effective if they are comprehensive at national, state or regional levels. There are far too many species to cover here, and they already feature in existing identification guides and scientific publications.

Scientists classify lizards according to common ancestry. This means how closely they are related, with allied species placed together in genera. It is traditional to present each genus and species in alphabetical order within their respective families but that approach has not been strictly followed here.

To the casual observer, what lizards look like and what they do is more important than their genetic affinities. Thus in this book lizards are clumped according to how they appear, the survival strategies they share, and even by factors affecting their distribution or how they relate to us. In many cases this logically follows their accepted classifications, but there are exceptions.

The many species of long-bodied, short-limbed burrowing skinks may not be related to each other but they have quite separately evolved similar extreme modifications that work well for them all, enabling them to slip through soil and compost. Likewise, a host of small skinks in dry climates face a common problem of dehydration, so unrelated species have

▲ Eastern Water Dragons *(Physignathus lesueurii)* thrive in the heart of many large towns and cities. Brisbane, Qld.

▲ The complex lattice pattern on the iris of the Spiny-tailed Gecko (*Strophurus krisalys*) helps conceal its eye. Longreach area, Qld.

▲ Robust Velvet Gecko (*Oedura robusta*). These geckos are equally at home on trees, rocks or house walls. Kurwongbah, Qld.

▲ Merten's Water Monitors (*Varanus mertensi*) inhabit waterways across northern Australia. Kununurra, WA.

independently been able to curb moisture loss by capping their eyes with fixed, clear windows. In this book, the lizards that have adopted these ways to exploit their respective environments have been placed together in two groups. Many of them look and act alike, and they share important physical and behavioural features.

On other occasions cousins that would normally appear separately are brought together. The blue-tongued skinks (*Tiliqua*) have much in common with their relatives the pink-tongued and she-oak skinks (*Cyclodomorphus*), including diet, defensive behaviour and general appearance. They are all closely related, yet alphabetically they sit astride the two ends of the skink family. In this book, their similarities draw them to adjoining text.

The geckos that have adapted easily to the human environment – 'house geckos' – have been treated together, while special categories have been created to deal with the challenges of dwelling in cold climates, isolation on mountain-tops and islands, and even the artful use of camouflage.

Of course, many species could easily be slotted into more than one category. There are burrowing skinks with long bodies that also have immovable spectacles covering their eyes to save moisture, while several of the earless dragons (*Tympanocryptis*) could justifiably be placed with a small number of extremely cryptic dragons. In these cases we have had to make a judgement on the most appropriate pigeon-hole.

Unlike birds, most lizards do not have widely accepted common names. Whether we like it or not, scientific names are our primary labels for these animals. Scientific names are applied throughout this book, and where common names are in use, they have been employed as well. For readers seeking information on particular lizards, you will do better to search them out using the thorough index rather than hunting for them alphabetically in the body of the book.

▲ A Masked Rock Skink (*Egernia margaretae personata*) and a Tawny Dragon (*Ctenophorus decresii*) bask together on a flat rock at Telowie Gorge, SA.

▲ The Pygmy Blue-tongue (*Tiliqua adelaidensis*) is the smallest of Australia's six blue-tongue species. Burra area, SA.

▲ Rainbow Skinks (*Carlia schmeltzii*) are aptly named because males display striking breeding colours. Chuwar, Qld.

Meet the skinks: an outrageous success story

▲ Bold-striped Robust Slider (*Lerista gerrardii*), one of many burrowing species with reduced limbs. Ninghan Station, WA.

If we measure success by variety of species and sheer abundance of animals, then skinks win the prize hands down. If success is defined by animals' abilities to exploit diverse environments, and modify their physical forms to entrench themselves within all available niches, skinks triumph again.

With evolution as the sculptor and skinks the clay, these versatile lizards have been stretched and compacted, shrunk, expanded and flattened. Their scales have been polished to a high gloss or carved into spines while their limbs and digits have been lengthened, or reduced and discarded. More than any other terrestrial vertebrates, the malleable skinks have submitted their bodies to extreme makeover.

As a result, nearly half of all Australian reptiles are skinks. About 400 species occupy virtually every habitat from coastal rocks to alpine summits, from rainforests to highly modified inner urban areas. They include huge ponderous gluttons, secretive miniature lizards smaller than many insects, worm-like troglodytes and racy desert speedsters complete with GT stripes. Then, of course, there are the common garden lizards familiar to us all.

▲ Red-throated Skink (*Acritoscincus platynotum*). Stanthorpe, Qld.

▲ Striped Skink (*Ctenotus iapetus*). Bullara Station, WA.

▼ Hosmer's Skink (*Egernia hosmeri*). Dajarra, Qld

▲ One of the largest Australian skinks, the Eastern Blue-tongue (*Tiliqua scincoides*) is a familiar resident of many backyards. Brisbane, Qld.

Typical skinks are smooth shiny lizards with four limbs and disposable tails. They feed on small insects that they seize in their jaws and swallow whole. Larger species often prey on smaller ones, while the biggest skinks tend to be omnivorous, leaning towards herbivorous. As they forage, skinks periodically flick out their broad flat tongues to seek airborne chemical cues.

Most skinks reproduce by laying soft, parchment-shelled eggs. The females of some species aggregate to lay their eggs communally. Sometimes these communal caches, containing several dozen eggs, represent the efforts of more than one species.

Some species of skinks produce very thin-shelled eggs that are retained in the body until the young are born fully developed. They are referred to as live-bearers. While this mode of reproduction occurs Australia-wide, it is particularly effective in cool climates where sun-warmed egg deposition sites are unreliable. It is a matter of benefits versus costs. The advantages of posturing and basking to warm the incubating embryos outweigh the peril that heavily pregnant females face from potentially greater predation risks.

Skinks are usually active by day, raising body temperatures by basking in the sun, and regulating them by shuttling between sun and shade. Yet some highly sensitive species shun direct sunlight, confining their activities to damp, dark places under rotting logs. Several are reported to expire from heat stress if held in the human hand.

There are few hard and fast rules with skinks. With nearly limitless variety, there is so much scope for species to buck the trends and break new ground (or burrow under it). But one thing certainly holds true: Skinks are an outrageous success story!

Lumbering giants: blue-tongues and the Shingleback

▲ Shingleback (*Tiliqua rugosa*). Perth, WA.

The slow-moving blue-tongues are some of the largest skinks in the world, reaching a head and body length of up to 30 centimetres. They are among Australia's most familiar and popular reptiles, and are frequently kept as pets. Generations of children have lovingly lavished a diet of mashed banana, meat and garden snails on their lizards.

Between them, the six Australian *Tiliqua* species live in virtually all Australian habitats, including home gardens within our capital cities. A seventh species occurs in New Guinea. They shelter under debris, in thick low vegetation and in the disused burrows of other animals. Blue-tongues are sun-loving lizards, often seen basking on garden paths. Unfortunately they are also fond of wandering onto roads, where traffic takes a heavy toll.

These plump, short-limbed lizards are omnivorous and generally greedy, consuming seasonal fruits, soft vegetation, flowers, birds' eggs, insects, molluscs and even treats like litters of baby mice if they find them.

Blue-tongues rely on bluff rather than speed to avoid being eaten. By gaping their pink mouths, protruding their flat, dark blue tongues and hissing they present a fearsome spectacle to any would-be predator. And if all else fails, those powerful jaws and blunt rounded teeth can clamp shut like a vice.

▲ Shingleback (*Tiliqua rugosa*). Quilpie, Qld.

All are live-bearers, with litters ranging from 18 for the Eastern Blue-tongue (*Tiliqua scincoides*) to one or two for the Shingleback (*T. rugosa*).

The **Shingleback** (*Tiliqua rugosa*) is a distinctive species with huge pine cone-like scales. It has a plethora of alternate names, so depending on where you live it may be a Stumpy-tail, Sleepy Lizard, Bobtail or Boggi. Heavily armoured, it moves slowly and ponderously through dry terrain over much of southern Australia. In some respects it can be considered Australia's answer to a terrestrial tortoise.

It probably consumes more vegetable material than other blue-tongues. Individuals live separate lives throughout the year, but seek out the same partner each season. During spring pairs can often be seen together, walking in tandem.

When an Eastern Brown Snake was struck and killed by a car near Burra, South Australia in 1992 a chain of events unfolded to solve one of Australia's great wildlife mysteries. The snake was dissected and found to contain a **Pygmy Blue-tongue** (*Tiliqua adelaidensis*), a species presumed to be extinct for over 30 years. Previous searches had all proven fruitless while populations had existed under our noses. They were overlooked because they live where no-one had dreamt of looking – in the vertical shafts of spider-holes on treeless grassy plains and slopes. These tiny blue-tongues have a head and body length of only 9 centimetres. They forage infrequently, bask near their burrows and seize invertebrates that stray near the burrow entrance. Studies have since revealed healthy populations of the Pygmy Blue-tongue within a limited area of South Australia.

▲ Pygmy Blue-tongue (*Tiliqua adelaidensis*). Burra, SA.

▲ Pygmy Blue-tongue (*Tiliqua adelaidensis*). Burra, SA.

▲ Eastern Blue-tongue (*Tiliqua scincoides*). Carseldine, Qld.

The **Eastern Blue-tongue** (*Tiliqua scincoides*) thrives in some of the busiest eastern Australian cities. These large lizards, with a head and body length of 30 centimetres or more, have been studied to investigate why they are so successful in the human environment. Females carrying young are very sedentary and less susceptible to danger than roving males; they produce large litters of young; the lizards rarely stray far from safe locations; they enjoy the artificial shelter sites we create, the gardens we grow with their attendant snails; and they grow rapidly and live for a long time. Captive Eastern Blue-tongues have been recorded to live for 30 years.

▲ The Centralian Blue-tongue (*Tiliqua multifasciata*) occurs in the Central Australian deserts. Ballera region, Qld.

Pink-tongue, she-oak skinks and slender blue-tongues

▲ Snails are a favourite food of the Pink-tongued Skink (*Cyclodomorphus gerrardii*). Mt Glorious, Qld.

The slender blue-tongues, she-oak skinks and the Pink-tongued Skink (*Cyclodomorphus*) are close relatives of the true blue-tongues but they are much more slender, longer tailed and more secretive. All are nocturnal or diurnal depending on temperature. They tend to occupy dense low vegetation, wriggling through thickets in a snake-like manner with the hind limbs tucked closely to the body.

Like their larger, more robust relatives, they are live-bearing skinks that employ their tongues to deter predators. But rather than just gape a cavernous mouth and present a broad flat tongue, these lizards often rear their bodies and rapidly flicker their tongues to resemble threatening snakes.

The **Pink-tongued Skink** (*Cyclodomorphus gerrardii*) occurs in moist, subtropical to tropical parts of eastern Australia. The tongues of juveniles are dark blue, and adults acquire their distinctive tongue colour as they grow. Two main colour forms occur within any population – those with and without bands. Juveniles are always boldly banded. The long tail is partially prehensile, allowing these semi-arboreal skinks to climb among vines and foliage.

Pink-tongued Skinks feed mainly on snails, cracking the shells with enlarged rounded rear teeth, manipulating the soft contents and discarding the shell fragments.

▲ Juvenile Pink-tongued Skinks (*Cyclodomorphus gerrardii*) are strikingly banded. Mt Glorious, Qld.

The three species of she-oak skinks are restricted to the south-east, particularly in temperate heaths. They can sometimes be seen basking among tussocks. The **Alpine She-oak Skink** (*Cyclodomorphus prealtus*) is confined to high altitude heaths, herb-fields and subalpine woodlands above 1500 metres. Its sensitive alpine habitat is under pressure from grazing, recreation and fires. It takes a long time for the montane landscapes to recover from disturbance, so the species is listed in Victoria as Endangered.

▶ Alpine She-oak Skink
(Cyclodomorphus prealtus).
Falls Creek, Vic.

23

The **Tasmanian She-oak Skink** (*Cyclodomorphus casuarinae*) is restricted to Tasmania but is widespread across the island.

The five species of slender blue-tongues are mainly found in seasonally dry to arid habitats.

The **Western Slender Blue-tongue** (*Cyclodomorphus celatus*) is common in sandy areas along the west coast, including coastal suburbs of Perth.

It is often found partly buried in loose sand under debris. Slender blue-tongues feed on a variety of invertebrates, and will also take smaller lizards.

▲ Tasmanian She-oak Skink (*Cyclodomorphus casuarinae*). Mt Field NP, Tas.

▲ Western Slender Blue-tongue (*Cyclodomorphus celatus*). Perth, WA.

Spiny, crevice, tunnelling and giant skinks: the egernias

▲ A Gidgee Skink (*Egernia stokesii*) uses its spiny scales to wedge itself into a split log. Aramac, Qld.

The 30 species of *Egernia* are a mixed bag of stoutly built, medium-sized to enormous skinks, ranging from a head and body length of about 9.5 to 30 centimetres. It is an extremely variable genus, with species tending to form clusters based on shared physical features and lifestyles, but between them they occur in virtually all habitats across the whole continent.

The smallest species are mainly insectivorous while larger egernias eat significant amounts of whatever vegetation is in season, including fruits, fungi and some foliage.

Most are active by day though some are mainly crepuscular and a few are nocturnal. All are live-bearers. *Egernia* includes Australia's most social lizards, and the young frequently live with parents and other adults in small family groups. They often share communal defecation sites. Encountering aggregations of droppings is a good indicator that a discreet colony of lizards is living nearby. All species seem to have a home range and permanently occupy some form of retreat, be it a burrow, log or rock crevice.

Spiny egernias

Four species of *Egernia* have spines on their bodies and tails. Most are associated with rocky habitats, and by inflating their bodies and arching their backs, they can wedge themselves so securely into rock cracks and crevices that there is little most predators can do to remove them.

▲ Cunningham's Skink (*Egernia cunninghami*). Girraween NP, Qld.

Some also extend into timbered areas where they inhabit hollow logs. All include distinctively coloured, geographically separated populations.

Cunningham's Skink (*Egernia cunninghami*) has a long, cylindrical tail with short spines. Distinctively coloured and patterned populations are spread along eastern Australia, north to the Carnarvon Range in central Queensland. The brightly marked animal shown here is typical of those living in the upland granites of the New England Tablelands in north-eastern New South Wales and adjacent Queensland.

Crevice egernias

Eight species of very muscular crevice egernias have flat heads and bodies, smooth to multi-keeled dorsal scales and generally dark colouration. Rock-inhabiting lizards occupy narrow crevices such as exfoliating slabs and rock clefts while those dwelling in trees live in hollows and behind loose bark.

The most widespread member of this group in eastern Australia is the **Tree Skink** (*Egernia striolata*). Despite the name, populations inhabiting rocks are as common as those in timber.

The **South-western Crevice Skink** (*Egernia napoleonis*) lives in woodlands, dry forests and outcrops of southern Western Australia. It also extends into heaths, living in the hollow stems of grass trees.

▲ Tree Skink (*Egernia striolata*). Mount Karong, Vic.

▼ South-western Crevice Skink (*Egernia napoleonis*). Eglinton area, WA. B. Maryan

Swamp egernias

The two species of swamp-dwelling egernias are confined to cool southern Australia, with one on either side of the continent. These extremely secretive, highly polished skinks bask discreetly in close proximity to thick low vegetation in marshy areas. They shelter under logs, often in saturated conditions, and both have been recorded to use crayfish burrows.

The **Western Swamp Skink** (*Egernia luctuosa*) lives in the south-west, between Perth and the Albany region. Unfortunately, wetlands destruction in the Perth area is threatening local populations.

▲ Western Swamp Skink (*Egernia luctuosa*). Lake Herdsman, WA.

▲ Eastern Swamp Skink (*Egernia coventryi*). Tooradin, Vic.

▼ Great Desert Skink or Tjakura (*Egernia kintorei*). Yulara area, NT.

▲ White's Skink (*Egernia whitii*), striped form. Freycinet NP, Tas.

The **Eastern Swamp Skink** (*Egernia coventryi*) is often associated with tea tree thickets and tidal salt marshes in the south-eastern corner of Australia.

Tunnelling egernias

The nine species of burrowing egernias are thick-set, smooth-scaled lizards, usually with short, deep heads. Most excavate simple to complex tunnels, often with one or more concealed escape exits, at the bases of shrubs in sandy deserts and heaths. The primary entrances to these tunnels are often indicated by loose, freshly excavated sand imprinted with footprints and tail dragmarks. Some cool temperate species in the south-east and south-west dig shallow tunnels under rocks.

The largest and rarest of this group is the **Great Desert Skink** or **Tjakura** (*Egernia kintorei*), with a head and body length of 24 centimetres. This lizard from the central and western deserts excavates complex burrow systems up to 1 metre deep and 4 metres long. A community of lizards share the accommodation and a nearby latrine. It is listed under Commonwealth legislation as Vulnerable.

White's Skink (*Egernia whitii*) digs simple tunnels under rocks and logs, sometimes merely creating a scrape to expand an existing cavity. It is the most southerly *Egernia*, and the only species to reach Tasmania. Most populations include two colour forms – striped and plain-backed lizards. In Tasmania, only the striped form occurs.

Large numbers of **Masked Rock Skinks** (*Egernia margaretae personata*) can often be seen perching at the entrances to deep crevices along rocky creek edges in the Flinders Ranges, South Australia. These lizards prefer homes that require little or no modification.

▲ White's Skink (*Egernia whitii*), plain-backed form. Stanthorpe NP, Qld.

▲ Masked Rock Skink (*Egernia margaretae personata*). Blinman Creek, Flinders Ranges, SA.

Giant egernias

Egernia includes some massive skinks. Species with a head and body length exceeding 18 centimetres occur in many parts of Australia. Large egernias tend to be very shy species, though in some popular national parks and picnic sites they can become familiar enough to accept handouts and scrounge for scraps.

King's Skink (*Egernia kingii*) is restricted to southern Western Australia. This lizard, with a head and body length of 24 centimetres, lives in a variety of habitats throughout the south-west, particularly on granite and coastal limestone. High population densities are found on some offshore islands.

If a massive, shiny black skink crashes through the undergrowth during a walk in subtropical rainforest, it is almost certainly a **Land Mullet** (*Egernia major*). This heavyweight, with a head and body length of 30 centimetres, is one of the largest skinks in the world. Land Mullets are usually very secretive but in some popular picnic grounds and walking trails they become habituated and easily approached. They live in hollow logs and in burrows dug into the earth-bound root system of fallen trees. Adults are patternless but juveniles, like the one shown here basking with its mother, are liberally marked with white spots

The **Yakka Skink** (*Egernia rugosa*) is mainly associated with Queensland's Brigalow Belt. Colonies occupy disused rabbit warrens and hollow logs. Much of their habitat has been cleared, fragmenting the populations. They are listed under Commonwealth legislation as Vulnerable.

▲ King's Skink (*Egernia kingii*). Carnac Island, WA. B. Maryan

▲ An adult Land Mullet (*Egernia major*) with a juvenile. Springbrook, Qld.

▲ Yakka Skink (*Egernia rugosa*), adult and juvenile basking. Alton, Qld.

In the backyard: eastern garden skinks

▲ Grass Skinks (*Lampropholis guichenoti*). Gerringong, NSW.

▼ Communal egg-cache, *Lampropholis* sp. Dandenong Ranges, Vic.

Few Australian home gardens are without their resident small brown skinks. In eastern Australia, the 11 *Lampropholis* species are alert, active, sun-loving skinks that favour well-watered habitats. We are familiar with those that bask together along our garden edges, skittering into rockeries, among shrubs and under leaf litter when disturbed. But other *Lampropholis* are much less obvious. Several are restricted to mountain-tops and others live only in rainforests.

Lampropholis are egg-layers that sometimes deposit their clutches in communal caches. Nests containing up to 100 eggs represent the efforts of many females that have come together at the same time. Sometimes the presence of older eggs indicates that the site has been used repeatedly over a number of seasons. We do not know what cues draw them or why they often lay together.

The **Grass Skink** (*Lampropholis guichenoti*) has adapted particularly well to our modified human environments. It thrives in parks and backyards throughout the south-eastern mainland. The skinks above are basking together in a home garden. The irregular dark vertebral stripe is typical of this species.

▲ Garden Skink (*Lampropholis delicata*). Kurwongbah, Qld.
▼ *Lampropholis mirabilis.* Magnetic Island, Qld.

Garden Skinks (*Lampropholis delicata*) extend along most of the east coast. They are among Australia's most successful reptile species but unfortunately they can also be invasive when introduced outside their natural range. They have been introduced to Hawaii, where there are no native lizards, and to Lord Howe Island and New Zealand, where they may have a negative affect on local skinks.

Lampropholis mirabilis inhabits boulders within pockets of rainforest and hoop-pines in a small area of north Queensland between Magnetic Island and Mt Elliott. It is an agile lizard, equally comfortable on vertical and horizontal surfaces and able to leap easily between boulders. It follows a common trend among skinks that forage on rock faces, having long limbs and digits and a flattened body.

Water, forest and barred-sided skinks

▲ Eastern Water Skink (*Eulamprus quoyii*). Girraween NP, Qld.

The 15 species of *Eulamprus* are moderate to large sun-loving lizards. They include voracious hunters that patrol within a defined home range, and 'sit and wait' predators that generally rest with the head and forebody protruding from a crevice or cavity. Most are associated with moist habitats, and all are live-bearers. They are restricted to eastern Australia.

Water skinks

Five species are popularly known as water skinks. The large, sleek, copper-coloured lizards are seldom far from water, patrolling fallen logs and rocks along creek banks and swamp edges. These alert lizards have no hesitation in taking a swim when pursued or even to capture prey. Although they are active foragers, they generally remain within their home ranges, using regular basking and shelter sites. Water skinks often include smaller lizards in their primarily insect-based diet.

The largest and most aquatic species, the **Eastern Water Skink** (*Eulamprus quoyii*), has a head and body length of 11 centimetres. It ranges from the Murray River to north-eastern Queensland. This species has adapted well to urban environments, as comfortable on a concrete culvert as on a stream-side log. The individual above entered a shallow pool and remained there for most of an afternoon, capturing insects attracted to the water.

The **Alpine Meadow Skink** (*Eulamprus koscuiskoi*) prefers high-altitude swamps with plenty of tussocks and sedges. During winter, when its habitat is covered by snow, it hibernates in tunnels under embedded rocks and logs. It is reported to be very aggressive to others of its own kind, perhaps because shorter seasons increase competition within populations.

Forest skinks

The five species of forest skinks are confined to rainforests, and each of three major forest blocks (north-eastern New South Wales to south-eastern Queensland; mid-eastern Queensland; and north-eastern Queensland) supports its own endemic species. Forest skinks are often confiding and easily approached as they perch on mossy logs beside walking trails. From these vantage points they can spy foraging invertebrates.

▲ Murray's Skink (*Eulamprus murrayi*). Lamington Plateau, Qld.

▲ Alpine Meadow Skink (*Eulamprus koscuiskoi*). New England NP, NSW.

▲ Lemon-barred Forest Skink (*Eulamprus amplus*). Mt Blackwood, Qld.

▲ Yellow-blotched Forest Skink (*Eulamprus tigrinus*). Lake Barrine, Qld.

▲ Greater Barred-sided Skink (*Eulamprus tenuis*). Southbrook, Qld.

Murray's Skink (*Eulamprus murrayi*) (page 35) has a distinctive dusting of fine pale blue spots on its flanks. It lives in subtropical rainforests of northern New South Wales and southern Queensland. Most populations appear stable but there is an alarming decline and possible localised extinctions in some northern parts of its range.

The **Lemon-barred Forest Skink** (*Eulamprus amplus*) from mid-eastern Queensland is closely associated with rocks as well as rotting logs and the buttresses of large figs. It can often be found sleeping at night on exposed boulders or tree roots. This attractive lizard is most common near sunlit edges of cascading rainforest streams

The **Yellow-blotched Forest Skink** (*Eulamprus tigrinus*) pictured here is perching in the disused tunnel of a large wood-boring insect. It is confined to the Wet Tropics region of north-eastern Queensland, extending from the lowlands to near the summits of the highest peaks.

Barred-sided skinks

There are five species of barred-sided skinks, which share a dark, deeply notched stripe along the flanks. In some species this is broken into a row of bars or blotches. They normally also feature narrow dark bands across their tails. They are common in rock outcrops, woodlands, tall moist eucalypt forests, occasionally in rainforests and in some urban areas from about Sydney northwards. Like the forest skinks they frequently bask motionless in semi-shaded situations, surveying their domains with just the head and fore body exposed.

The **Greater Barred-sided Skink** (*Eulamprus tenuis*) sports a dark lateral stripe that is usually broken into blotches. It is common in garden rockeries and frequently enters houses in Sydney and Brisbane. For this lizard, the view from under a washing machine is just as fine as that from a rock outcrop or forest log.

37

Cool customers: sun-loving skinks in southern climates

▲ The Metallic Skink (*Niveoscincus metallicus*) is abundant in Tasmania and south-eastern Victoria. Tooradin, Vic.

The cool temperate regions, including the New South Wales tablelands, Australian alps and Tasmania support fewer reptile species than warmer climates, but some skinks are able to exploit the sometimes meagre sunshine to good effect. They face obvious challenges in keeping their bodies warm, yet the lizards that occur in cold climates often thrive in large numbers. They bask on the sheltered, sunny sides of rocks or among tussock-grass clumps. Thanks to a combination of dark colouration, flattening their bodies and selecting ideal basking sites, they can raise their temperature above that of the air around them. Some skinks even bask together to form a communal, heat-absorbent mat. Some can forage, seize prey and dash swiftly for cover within one metre of thawing snow. Most of these cool customers give birth to live young.

Snow skinks

The eight species of snow skinks (*Niveoscincus*) are confined to Tasmania and the south-eastern mainland. They are typically dark coppery brown, though some are attractively patterned with pale spots or flecks.

▲ Pedra Branca Skink (*Niveoscincus palfreymani*). Pedra Branca Rock, Tas. S. Donnellan

The **Pedra Branca Skink** (*Niveoscincus palfreymani*) is one of the world's most restricted reptiles, clinging to existence on a remote 1.4 hectare rock devoid of vegetation 26 kilometres south of the Tasmanian mainland. There it feeds on flies, small coastal crustaceans and fish scraps scavenged from the local seabird colony.

Tussock skinks

The six tussock skinks (*Pseudemoia*) occur in areas dominated by tussock grasses. They are largely restricted to the south-east of Australia, with one species occurring along the Great Australian Bight. Most are patterned with narrow dark stripes, and breeding males of some species develop narrow red stripes along the flanks. All give birth to live young.

Grasslands Tussock Skink (*Pseudemoia pagenstecheri*). The lizard on page 40 is foraging among thawing snow. In southern parts of its range, the species is common among tussocks in open lowland habitats. Northwards into New South Wales, populations become increasingly dependant on higher altitudes where cool conditions prevail.

The **Glossy Tussock Skink** (*Pseudemoia rawlinsoni*) is a secretive inhabitant of dense low vegetation, often associated with swamps and the edges of watercourses. It basks discreetly among thick tussock foliage. The lizard pictured on page 40 has closed its eyes yet it still has some vision. Like many small skinks its moveable lower eyelids contain small transparent windows.

39

▲ Grasslands Tussock Skink (*Pseudemoia pagenstecheri*). Polblue Swamp, Barrington Tops, NSW.

▲ Glossy Tussock Skink (*Pseudemoia rawlinsoni*). Warneet, Vic.

Old snake-eyes: small skinks conserving water in dry climates

▲ Elegant Snake-eyed Skink (*Cryptoblepharus pulcher*), showing the fixed, clear spectacle over its eye Kurwongbah, Qld.

Small skinks that are active by day in hot dry climates or low-moisture habitats run a serious risk of dehydration. A potential source of evaporative loss is the surface of the eye. Many of these skinks cap the eye with a fixed, clear spectacle. In effect, the lower eyelid is transparent and fused shut.

Covering the eye with an immovable transparent window has evolved independently among small skinks worldwide. More than 80 Australian species exhibit the condition. At one time, all skinks that adopted this water-saving strategy were lumped together in one genus, *Ablepharus*. The name often appears in old Australian reference books and was applied to unrelated species ranging from central Victoria to central Europe.

Morethias

The eight species of *Morethia* are active sun-loving lizards, common in dry habitats across Australia. They are often the most conspicuous small terrestrial skinks, foraging in open areas but rarely far from the cover of leaf litter or low vegetation. Breeding males of most *Morethia* species develop red throat flushes.

The **Pale-flecked Morethia** (*Morethia lineoocellata*) lives along the west coast. It belongs to a group of generally weakly patterned southern species that occupy open woodlands and shrublands. Juveniles often have a reddish tail-flush that fades on adults.

▲ Pale-flecked Morethia (*Morethia lineoocellata*). Rottnest Island, WA.

Boulenger's Morethia (*Morethia boulengeri*) has a distinctive tan back and prominent white stripe along the flanks. It is one of the most abundant skinks across vast tracts of seasonally dry southern Australia. It favours firm soils supporting a mix of woodlands and low shrubs with crisp, dry leaf litter.

The **Eastern Fire-tailed Morethia** (*Morethia taeniopleura*) lives along the north-eastern coast and ranges. It belongs to a more northern and central group, often associated with rocky areas. Members of this group, often referred to as fire-tailed skinks, are sharply marked with prominent pale stripes. The tail, usually an intense red at all ages, is sinuously waved as they forage.

▲ Boulenger's Morethia (*Morethia boulengeri*). Meandarra, Qld.

▼ Eastern Fire-tailed Morethia (*Morethia taeniopleura*). Wivenhoe, Qld.

Snake-eyed skinks

The 23 species of snake-eyed skinks (*Cryptoblepharus*) occur virtually Australia-wide, excepting the south-eastern mainland and Tasmania. They extend from the deserts well into areas with high rainfall, but because they mainly dwell on exposed vertical surfaces, the sites they occupy are relatively moisture free. Also known as wall skinks and fence skinks, these flat-bodied, long-limbed lizards are extremely swift, shinning with gravity-defying ease over rocks, tree trunks, buildings and fences.

Despite their abundance within urban environments very few egg clutches have been recorded, though some are known to lay communally. Some species are recorded to augment their hunting activities by staking out ant trails to rob the insects of the cargo they carry. They are excellent dispersers and seafarers, with a range of species established on islands between Africa and the western Pacific.

The **Elegant Snake-eyed Skink** (*Cryptoblepharus pulcher*) is a regular city slicker, thriving on walls and fences from south-eastern New South Wales to north-eastern Queensland. The courting male pictured has grasped a female, hoping to mate.

The **Coastal Snake-eyed Skink** (*Cryptoblepharus litoralis*) rarely strays more than 100 metres inland. It clings so closely to the rocky coasts of north-eastern Queensland and the Northern Territory that it dodges incoming waves and forages on the wet rocks in their wake.

▲ Elegant Snake-eyed Skink (*Cryptoblepharus pulcher*). Kurwongbah, Qld.

▲ Coastal Snake-eyed Skink (*Cryptoblepharus litoralis*). Prince of Wales Island, Qld.

▲ Agile Snake-eyed Skink (*Cryptoblepharus zoticus*). Lawn Hill NP, Qld.

The **Agile Snake-eyed Skink** (*Cryptoblepharus zoticus*) is a long-limbed, flat-bodied rock specialist, dwelling on sandstone outcrops and escarpments of north-western Queensland and adjacent Northern Territory.

Dwarf skinks

The eight species of dwarf skinks (*Menetia*) are swift and alert but extremely secretive, rarely venturing beyond the cover of leaf litter. They occupy the cavities between fallen leaves, popping their heads up among dry leaf litter, then vanishing to re-appear somewhere nearby. These are some of Australia's smallest reptiles, with total head and body lengths as little as 2.5 centimetres. They hunt tiny insects, but in a reversal of the typical 'lizards prey on invertebrates' scenario, these diminutive skinks must certainly feature regularly in the diets of a host of spiders, centipedes and carab beetles. They occur across most of Australia, with their highest diversity in the dry tropics.

The juvenile **Common Dwarf Skink** (*Menetia greyii*) is curled up on a human thumbnail with plenty of room to spare. It is hard to imagine that such a tiny lizard contains all the same vertebrate features as we do – a skeleton, lungs, perfectly formed digits and a beating heart. This is the most widespread of the dwarf skinks, occurring in dry areas of all mainland states.

Sadlier's Dwarf Skink (*Menetia sadlieri*) has a much more restricted distribution. It is known only from Magnetic Island near Townsville. There is currently heavy development pressure on lowland parts of the island, which may threaten this tiny lizard. The head and body length is less than 3 centimetres.

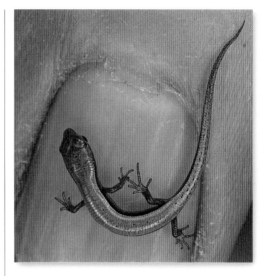

▲ Common Dwarf Skink (*Menetia greyii*), juvenile. Collinsville, Qld.

▲ Sadlier's Dwarf Skink (*Menetia sadlieri*). Magnetic Island, Qld.

The underground movement: when legs get in the way

▲ This Keeled Slider (*Lerista planiventralis*) has a streamlined snout. Bullara Station, WA.

▲ The snake skinks, like *Ophioscincus ophioscincus*, have rounded snouts to push through compost. Mt Glorious, Qld.

About one third of Australian skinks are burrowing species. They illustrate just how versatile the skink family can be when it comes to extreme makeover. To meet the challenges of a life spent wriggling through the soil these skinks have evolved elongated bodies for greater flexibility, while their snouts may be either streamlined to help slip through dry, loose sand or reinforced with a tough cuticle to push through damp compost.

The most significant changes are the limbs. Beneath the ground these just get in the way so their widely-spaced appendages have become shorter. Sometimes digits have been lost, while many species have completely forfeited their forelimbs but retain hind limbs to push them through sand. Others are left with mere stubs and several have lost all external traces of legs. Exposed ear-openings can also be a problem so these are often reduced in size or covered by scales.

The underground movement has obviously been hugely attractive to skinks. Burrowing species with long bodies and reduced or missing limbs occur in a range of Australian habitats from wind-blown, sandy deserts to high-altitude tropical rainforests. Several unrelated groups in Australia and overseas have independently adopted parallel lifestyles and evolved similar physical adaptations to burrow. In the process, some distant relatives have come to closely resemble each other. To the layperson, near-limbless skinks wriggling through dry sandy swales in far-away Madagascar and Africa look outwardly identical to some Australian burrowing skinks. They have evolved convergently to exploit similarly structured environments.

Sliders

The largest burrowing group are the sliders (*Lerista*). About 90 species exhibit the most complete sequence of burrowing modifications within one genus. Those that dwell under surface debris and in leaf litter are relatively unspecialised lizards with four short limbs, each with three to five digits. Species living in upper soil profiles show a progressive loss of fingers and toes, while limbs may be reduced to stubs. Some are streamlined, shovel-snouted skinks with little or no indication of limbs, able to swim through loose sand like fish through water.

Sliders are found throughout Australia, but they are most successful in dry habitats such as deserts and tropical northern woodlands. Some species thrive in dry conditions because they have capped the eye with a clear spectacle to conserve moisture.

Along the west coast, accumulated leaf litter under each acacia bush can support a community of *Lerista* species, while throughout the interior, red desert sands are criss-crossed with intricate meandering *Lerista* tracks. The least specialised *Lerista* – those that are most conventionally lizard-like with well-developed limbs – live under rocks and logs on the southern mainland and northern Tasmania.

Lerista apoda is completely limbless and has a flat, acutely shovel-shaped snout with its small eyes set beneath a transparent scale. It is restricted to the Dampier Peninsula in north-western Australia, where it creates wriggling tracks across the surface of pale coastal sand dunes.

Wilkins' Slider (*Lerista wilkinsi*) has no trace of a forelimb and only two toes on the hind limb. It burrows in fine sand beneath leaf litter and rocks in a small area of outcrops and gorges in the dry north-eastern interior of Queensland. The reddish tail flush is commonly seen on many immature *Lerista*.

▲ *Lerista apoda*. Broome, WA. D. Knowles

▼ Wilkins' Slider (*Lerista wilkinsi*). Warrigal Range, Qld.

Edwards' Slider (*Lerista edwardsae*) has the forelimb reduced to a simple stump and the short hind limb has only two toes. It inhabits loose sand under leaf litter in semi-arid shrublands and mallee woodlands across southern South Australia.

The **Keeled Slider** (*Lerista planiventralis*) has two fingers and three toes. Its wedge-shaped snout and the angular ridge along each side of its belly allow it to swim through loose sand on plains and desert dune-crests along the west coast of Western Australia.

The **Blue-tailed Slider** (*Lerista zietsi*) has four fingers and toes. It lives in soil under rocks among the spectacular landscape of deep gorges in Karijini National Park in the Pilbara region of Western Australia.

Bougainville's Skink (*Lerista bougainvillii*) has well-developed limbs and a full complement of digits. It is not as strongly adapted for burrowing as most of its relatives. This is also the most southerly member of this largely arid-adapted group, extending well into cool temperate areas including northern Tasmania.

▲ Edwards' Slider (*Lerista edwardsae*). Port Germein, SA.

▲ Keeled Slider (*Lerista planiventralis*), Bullara Station, WA.

▼ Blue-tailed Slider (*Lerista zietsi*). Hamersley Range, WA.

▲ Bougainville's Skink (*Lerista bougainvillii*). Telowie Gorge, SA.

▲ Verreaux's Skink (*Anomalopus verreauxii*), adult. Wivenhoe, Qld.

▼ Verreaux's Skink, juveniles. Brisbane, Qld.

Worm skinks

Worm skinks (*Anomalopus*) include three very short-limbed and four completely limbless lizards. Those species with limbs occur mainly in woodlands and open paddocks in mid-eastern Australia. They tend to live in heavy, loose soil and compost, often beneath slightly embedded logs and rocks.

Limbed worm skinks contradict a worldwide trend among burrowing lizards. In virtually all other cases, when limbs are reduced and digits lost, it is the forelimbs and fingers that go first. This means burrowing skinks have fewer fingers than toes, or their fingers and toes are equal in number, and their forelimbs are also generally shorter than their hind limbs. Yet the worm skinks have more fingers than toes. They retain a degree of functional use of their front legs while the hind limbs are nearly worthless for locomotion. The reason is unclear, but they may use their front legs to help pull themselves through soil. Their unique approach to limb reduction sets them apart from hundreds of other burrowing lizards around the world.

Verreaux's Skink (*Anomalopus verreauxii*) has three stubby fingers on the forelimb and a simple stump for a hind limb. This large species has a head and body length of over 18 centimetres. The bluish colour of the adult in the photograph indicates it is about to shed its skin. Juvenile Verreaux's Skinks have a prominent yellow collar across the neck. This darkens with age and virtually disappears on large adults.

Limbless worm skinks occur from Cape York to just north of Sydney, in habitats ranging from heaths to vine thickets. **Anomalopus brevicollis** lives in mid-eastern Queensland. It shelters in loose soil under rocks and leaf litter, vanishing rapidly into the substrate if disturbed.

▲ *Anomalopus brevicollis*. Cracow, Qld.

▼ Yolk-bellied Snake Skink (*Ophioscincus ophioscincus*). Mt Glorious, Qld.

Snake skinks

There are three species of snake skinks (*Ophioscincus*), confined to moist subtropical areas between northern New South Wales and southern Queensland. They have tiny eyes, relatively thick bodies with blunt-tipped tails, and their rounded snouts are protected from abrasion by a waxy cuticle.

The **Yolk-bellied Snake Skink** (*Ophioscincus ophioscincus*), named for its bright-coloured underside, is one of two completely limbless species. It lives among compost and damp soil in rainforests and vine thickets in south-eastern Queensland. If exposed to dry conditions it quickly desiccates.

The **Short-limbed Snake Skink** (*Ophioscincus truncatus*) has four limbs, but they are just minute stumps. It occurs in rainforests from northern New South Wales to southern Queensland, with outlying populations in heaths and forests on the sandy islands of Moreton Bay.

▲ Short-limbed Snake Skink (*Ophioscincus truncatus*). North Stradbroke Island, Qld.

Earless skinks

The five species of earless skinks (*Hemiergis*) are glossy brown above and bright yellow to orange below with rounded snouts, long thick tails and four very short, widely-spaced limbs with 2–5 fingers and toes. They extend across southern Australia, from the south-west (where they are secretive but very common garden lizards) to the Northern Tablelands of New South Wales. All live in soft sand or soil under mats of vegetation or semi-embedded logs and rocks. They are sensitive to high temperatures and moisture loss, yet two species live in semi-arid spinifex and open shrubland communities. Within these harsh environments they seek humid, sheltered retreats.

The **Four-toed Earless Skink** (*Hemiergis peronii*) extends between western Victoria and the south-western corner of Western Australia.

▶ Four-toed Earless Skink (*Hemiergis peronii*). Kangaroo Island, SA.

▲ Three-toed Snake-toothed Skink (*Coeranoscincus reticulatus*). Cunningham's Gap, Qld.

▲ Fraser Island Sand Skink *(Coggeria naufragus)*. Fraser Island, Qld.

▼ Three-toed Skink (*Saiphos equalis*). Girraween NP, Qld.

Snake-toothed skinks

There are two snake-toothed skinks (*Coeranoscincus*). A short-limbed and a limbless species occur in widely separated subtropical and tropical rainforests. They live under damp leaf litter and in rotting logs. Their pointed, recurved teeth are quite unlike the peg-like structures of other skinks, and probably assist in grasping moist slippery earthworms. Adults are mainly patternless and drably coloured but juveniles are boldly adorned with bands or stripes. The **Three-toed Snake-toothed Skink** (*Coeranoscincus reticulatus*) from southern Queensland and northern New South Wales has four limbs, each with three short digits.

Fraser Island Sand Skink

This unusual species, *Coggeria naufragus*, has no close relatives and is the sole member of its genus. It has a wedge-shaped snout, four limbs each with three short digits, and a head and body length over 12 centimetres. It is widespread on Fraser Island, Queensland, occupying a wide range of habitats from closed rainforest to heath. Despite numerous searches, there are no records from similar habitats on the adjacent mainland. They are extremely secretive and rarely encountered, yet fauna surveys involving pit-trapping have revealed them to be common.

Three-toed Skink

The **Three-toed Skink** (*Saiphos equalis*) lives in moist, loose soil under rocks, logs and leaf litter between south-eastern New South Wales and south-eastern Queensland. It has a bright yellow to orange belly and chest, merging to black under the long thick tail, and short, widely-spaced limbs with three digits. It gives birth to live young, though some populations lay eggs containing fully-formed embryos that hatch about a week later.

Lurking in moist compost: calyptotis and mulch skinks

▲ Garden Calyptotis (*Calyptotis scutirostrum*). Mt Glorious, Qld.

Mulch and calyptotis skinks are glossy-scaled lizards with short, well-developed limbs and moderate to very elongate bodies. Some are modified in much the same way as many burrowing skinks, though they have a full complement of five digits on each limb and tend to live in cavities under logs and in decomposing vegetation, rather than pushing their way through loose soil. They require slightly moist conditions. All species are egg-layers.

Calyptotis

The five species of *Calyptotis* usually have yellow bellies and pink beneath their tails. They are restricted to eastern Australia between north-eastern New South Wales and north-eastern Queensland. They are generally most abundant (sometimes extremely so) in relatively open, timbered areas where

numerous logs and plentiful leaf litter offer slightly moist shelter. These secretive skinks are not normally encountered active in exposed areas by day except during overcast humid weather.

The **Garden Calyptotis** (*Calyptotis scutirostrum*) is a common lizard in suburban Brisbane. Its pattern, typical of most *Calyptotis*, includes a glossy brown back, a dark streak along the upper flanks, and usually narrow dark lines down its back.

The **Thornton Peak Calyptotis** (*Calyptotis thorntonensis*) has a very restricted distribution on rocky slopes under thick closed rainforest between about 600 and 700 metres altitude on Thornton Peak, in the Wet Tropics of north Queensland. It is isolated in a cool montane habitat

▲ Thornton Peak Calyptotis (*Calyptotis thorntonensis*). Thornton Peak, Qld.

surrounded by warmer lowlands, and is reported to be extremely heat-sensitive, succumbing to the temperature of a human hand.

Mulch skinks

The 14 species of mulch skinks (*Glaphyromorphus*) are more variable than *Calyptotis*, ranging from extremely elongate species with short, very widely spaced limbs that live under logs and leaf litter, to lizards of more moderate proportions that tend to be crepuscular and often forage in open areas near thick low vegetation. They are mainly confined to northern Australia.

The **Orange-sided Mulch Skink** (*Glaphyromorphus douglasi*) occupies moist shelter sites across the north of the Northern Territory. It is often associated with compost heaps in suburban Darwin, and is commonly seen active at dusk. At one picnic site near Darwin, two lizards were observed engaged in a bout of tug-of-war with a discarded chop bone.

▲ Orange-sided Mulch Skink (*Glaphyromorphus douglasi*). Darwin, NT.

▲ Mjoberg's Skink (*Glaphyromorphus mjobergi*). Ravenshoe area, Qld.

◄ Black-tailed Mulch Skink (*Glaphyromorphus nigricaudis*). Iron Range, Qld.

Mjoberg's Skink (*Glaphyromorphus mjobergi*) is a very secretive inhabitant of rotting logs in cool, dense upland rainforest above about 600 metres in the Wet Tropics of northern Queensland.

The **Black-tailed Mulch Skink** (*Glaphyromorphus nigricaudis*) is a widespread species of tropical forests and woodlands, ranging from north-eastern Queensland to north-eastern Northern Territory and southern New Guinea. It emerges to forage at dusk.

▲ Black-tailed Mulch Skink. Tip of Cape York, Qld.

In muted tones: shade skinks

▲ Weasel Skink (*Saproscincus mustelinus*). Hunters Hill, NSW.

Shade skinks (*Saproscincus*) shun bright sunlight. They live in rainforests and moist temperate forests, where they can often be seen foraging on damp leaf litter or basking in dappled sun on logs and rocks beside tracks. In some areas, shade skinks are among the most frequently seen lizards as they can be very confiding and easy to approach. Others are highly secretive and seldom encountered active. They are not particularly swift.

The 11 species are all confined to eastern Australia. They have smooth, matt-textured scales, usually a pale dash on the rear base of the hind limb, and often a pair of rusty stripes along the tail. Some species are extremely variable, with a proportion of females exhibiting bold white stripes along black flanks. They range in size from a head and body length of over 6 centimetres, to little more than 3 centimetres.

Shade skinks are egg-layers. Some small northern species have a fixed clutch size of two, while at least one southern species lays up to seven eggs, often deposited communally.

The **Weasel Skink** (*Saproscincus mustelinus*) is the most southerly shade skink, occurring between northern New South Wales and southern Victoria. It is common in gardens around Sydney and Melbourne but is very secretive, generally avoids direct sunlight and tends to hug closely to rank thickets. Weasel Skinks are mainly active at dusk and rarely venture beyond the security of thick, low vegetation. They can easily be identified by the very long russet tail and the pale dash behind the eye. Weasel Skinks have been recorded to lay their eggs communally with garden skinks (*Lampropholis* species).

Rose's Skink (*Saproscincus rosei*) is one of Australia's most variable lizards. There are geographical and sexual differences, as the females within some populations can differ markedly from males. Rose's Skinks occur in rainforests and rainforest edges along eastern ranges of northern New South Wales and southern Queensland. The brightly coloured lizard from the Lamington Plateau pictured below exhibits a female colouration that occurs at low frequency in that region. Plainer individuals, like the skink above from Mt Glorious, are common across the species' range. The orange patches behind the lizards' forelimbs are clusters of mites.

▲ Rose's Skink (*Saproscincus rosei*), common plain colouration. Mt Glorious, Qld.

▼ Rose's Skink (*Saproscincus rosei*), uncommon female colouration. Lamington Plateau, Qld.

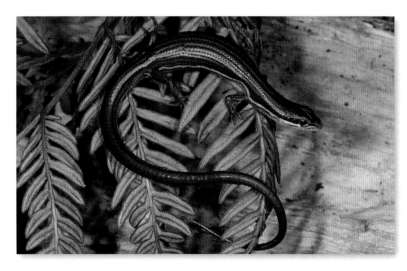

◄▼ Dwarf Weasel Skink (*Saproscincus oriarus*). Broken Head, NSW.

The **Dwarf Weasel Skink** (*Saproscincus oriarus*) is restricted to a narrow coastal strip of swamps, paperbark thickets, rainforests and gardens in northern New South Wales and southern Queensland. This very slender, long-tailed lizard always has a narrow pale stripe along its flanks.

For visitors trekking the walking trails in north Queensland's Wet Tropics, **Saproscincus basiliscus** is a common sight foraging in shade and dappled sun. It is a relatively confiding shade skink that can easily be approached with care.

▶ *Saproscincus basiliscus.* Curtain Fig NP, Qld.

Elfin shade-lovers: Maccoy's and Elf skinks

▲ Maccoy's Skink (*Anepischetosia maccoyi*). Illawarra region, NSW.

Two very small, secretive skinks live under logs and leaf litter in forests of eastern Australia. Both have fine pointed snouts. Their large eyes, translucent scales and small size tie them to permanently damp habitats. If exposed to warm dry air they would quickly desiccate.

Maccoy's Skink (*Anepischetosia maccoyi*) has short, widely-spaced limbs, a full complement of five digits, and an orange-yellow belly. It inhabits wet gullies and moist forests of the south-east, with populations scattered from the Grampians and Otway Ranges in Victoria to the highlands and Illawarra regions of New South Wales. Maccoy's Skink is extremely heat sensitive, with a preferred body temperature of only about 21° Celsius. It is an egg-layer,

recorded to sometimes deposit its clutches communally with Garden Skinks (*Lampropholis delicata*). The head and body length is 5 centimetres.

▼ Maccoy's Skink (*Anepischetosia maccoyi*). Illawarra region, NSW.

▶ Elf Skink (*Eroticoscincus graciloides*). Boreen Point, Qld.

The **Elf Skink** (*Eroticoscincus graciloides*) has four fingers and five toes, an iridescent sheen, and reddish brown stripes along the tail. It is confined to south-eastern Queensland, occurring in rainforests, tall moist eucalypt forests and in damp depressions within drier open forests, particularly near creeks. Elf Skinks are rarely encountered foraging, though they are sometimes seen after rain and during overcast weather. The head and body length is only 3 centimetres.

▲ Elf Skink (*Eroticoscincus graciloides*). Buderim, Qld.

The rough-backed skinks

▲ Prickly Forest Skink (*Gnypetoscincus queenslandiae*). Lake Barrine, Qld.

Two Australian skinks have extremely distinctive, granular and keeled scales, creating an almost crocodile-like effect. They are confined to closed forest habitats in Queensland, but their habitats and lifestyles are strikingly different. Both are highly secretive species that are rarely encountered active. They are live-bearers.

Prickly Forest Skink

Prickly Forest Skinks (*Gnypetoscincus queenslandiae*) live under rotting logs in rainforests of the Wet Tropics in north-eastern Queensland. They emerge infrequently and probably at night, so appear to live within a perpetually dark and damp environment. Prickly Forest Skinks are often locally abundant. In some upland rainforests, virtually every suitable rotting log harbours a resident skink.

The granular scales are believed to assist in keeping the skin evenly moist by dispersing water via capillaries along scale edges. On most other skinks, a droplet of water placed on their overlapping scales will be repelled and hold its shape, but on the Prickly Forest Skink it is immediately absorbed and spread across the surface like water on blotting paper.

▲ Prickly Forest Skink (*Gnypetoscincus queenslandiae*). Wooroonooran NP, Qld.

Nangur Spiny Skink

In 1992 researchers conducting a fauna survey removed an unusual lizard from a burrow in a dense vine thicket near Murgon, 250 kilometres north-west of Brisbane. The thickset skink had spiny, granular body scales and rugose head shields. It turned out they had discovered a new genus and species just a few hours drive from a major capital city.

The **Nangur Spiny Skink** (*Nangura spinosa*) is known from just two small populations about 60 kilometres apart. It is restricted to semi-evergreen vine thickets on heavy dark soils. Burrows are excavated in sloping ground at the bases of tree trunks, rocks and exposed roots. From vantage points at the burrow entrances, the lizards rest with the head and forebody exposed. Passing invertebrates are seized and the lizards return to the burrow entrances to feed.

Juveniles cohabit the burrow with their mother, probably remaining with her for many months. When she emerges from the burrow it is not uncommon for juveniles to actually appear on her back.

We do not know why Nangur Spiny Skinks have their unusual scale textures, and whether they disperse moisture like the Prickly Forest Skink. When they enter their burrows they often block the entrances with their thick tails, so the rough textures may help prevent predators from removing the lizards.

Most of the original semi-evergreen vine thickets of south-eastern Queensland have been cleared. Only relict, isolated pockets remain. Despite intensive searches, no additional populations of Nangur Spiny Skinks have been found. They may have been scarce and patchy prior to land clearing, but the combined adult population at both sites is now estimated at just 200. In late 2007, the Federal Government formally listed *Nangura spinosa* as Critically Endangered.

▼ A Nangur Spiny Skink Nangura Spinosa feeds near its burrow entrance. Nangur State Forest, Qld.

▲▼ Nangur Spiny Skink (*Nangura spinosa*). The lizard below is at its burrow entrance. Nangur State Forest, Qld.

It pays to advertise: rainbow skinks

▲ Blue-throated Rainbow Skink (*Carlia rhomboidalis*). Magnetic Island, Qld.

Rainbow skinks (*Carlia*) are well named. For these active sun-loving skinks, visual displays such as colour and tail-waving are important means of communication. Males of many species of rainbow skinks display breeding colours that rival some of the more spectacular dragons. They feature combinations of red stripes along the flanks, red or blue washes over the throat or orange flushes over the hips and tail. All species have four fingers and five toes.

The 32 named species extend across northern Australia, down the east coast to central Victoria. Diversity is highest in dry woodlands and outcrops of north-eastern Queensland. Most *Carlia* are active among sunlit leaf litter, moving swiftly over and through the leaves in search of arthropod prey. They are often the most abundant and conspicuous lizards in northern tropical woodlands, where the dry leaf litter along some creek margins seems literally alive with them.

A number of north-eastern species are exclusively rock-inhabiting. They have evolved flatter bodies with longer limbs and digits, enabling them to forage with ease over boulder surfaces. These rock-inhabiting *Carlia* generally lack breeding colours but exhibit conspicuous tail-waving and head-bobbing displays. All *Carlia* are egg-layers, producing fixed clutches of two eggs.

The **Blue-throated Rainbow Skink** (*Carlia rhomboidalis*) lives along sunny edges and clearings in rainforests of tropical mid-eastern Queensland. Most other rainbow skinks prefer drier habitats. Males seasonally acquire brilliant blue and pink throats, and indications of these colours can even be seen on some juveniles.

The **Rough Brown Rainbow Skink** (*Carlia johnstonei*) hunts in the leaf litter among gorges and escarpments of the Kimberley region, Western Australia. The skink pictured has captured a March Fly.

The males and females of the **Lined Rainbow Skink** (*Carlia jarnoldae*) could be mistaken for different species. The orange flush and bright blue spots on the male's flanks contrast sharply with the simple white stripe of the female. Lined Rainbow Skinks live in open forests, mainly on stony soils, in north-eastern Queensland.

▲ Rough Brown Rainbow Skink (*Carlia johnstonei*). Manning Creek, WA.

◄ Lined Rainbow Skink (*Carlia jarnoldae*), male. Petford, Qld.

▼ Lined Rainbow Skink (*Carlia jarnoldae*), female. Mt Elliot, Qld.

◀ Black Mountain Skink (*Carlia scirtetis*). Black Mountain, Qld.

The **Black Mountain Skink** (*Carlia scirtetis*) lives on just one mountain range, a surreal landscape of tumbled black boulders near Cooktown, north Queensland. The black lizards, with a head and body length of just over 6 centimetres, are extremely alert and agile, able to leap with ease between boulders. They also exhibit a degree of 'trepidatious curiosity' towards human observers.

The **Closed-litter Rainbow Skink** (*Carlia longipes*) below is a male in breeding condition. This attractive species, with a head and body length of over 6 centimetres, is the most conspicuous skink in the sunny clearings of Cape York's monsoon rainforests.

▲ Closed-litter Rainbow Skink (*Carlia longipes*). Iron Range, Qld.

Greased lightning with GT stripes: striped skinks

▲ *Ctenotus arcanus*, showing comb-like lobules in front of the ear opening. Mt Mee State Forest, Qld.

To the non-specialist eye, the more than 90 species of striped skinks (*Ctenotus*) are a confusing group of similarly marked lizards. Indeed, some continue to pose identification problems for trained herpetologists.

Striped skinks occupy a range of dry open habitats across Australia, except Tasmania. They are generally absent from closed forests and wetlands. Striped skinks occur within most Australian cities, but diversity is greatest in arid zones, particularly the central and western sandy deserts, and in dry tropical northern areas. Some species have ranges covering more than half the continent while others are known from single localities or small isolated habitats.

Most species have enlarged comb-like lobules along the front of the ear-opening and all have well-developed limbs with five fingers and toes. Their patterns range from simple (unbroken dark and pale stripes) to complex (a mix of stripes and rows of spots). A few species are spotted and several are plain.

Ctenotus are extremely swift, alert, sun-loving lizards that lay eggs and feed mainly on small invertebrates. Despite this broad similarity, physical and behavioural differences allow up to seven or more species to share a single desert location. Some are mainly active in the morning, others in the afternoon. Open-foragers with proportionally longer limbs make regular forays across bare ground while shorter-limbed skulkers rarely stray far from cover. Most are opportunistic feeders that devour any suitably-sized prey, but several appear to specialise on termites. And size variation is such that food eaten by some is unavailable to others. For example, *Ctenotus grandis titan* (with a head and body length of over 12 centimetres) could easily devour *C. rufescens* (just over 4 centimetres). *Ctenotus* exhibits classic examples of niche partitioning within ecological communities and the radiation of species across a drying continent.

Ctenotus agrestis has plain stripes, although one pale stripe along the upper flank is usually broken into dashes. It lives in a small area of central Queensland, where open grassland grows on cracking clay soil.

Ctenotus allotropis has a complex pattern of stripes and rows of pale spots. It occurs in dry open woodlands on sandy and loam soils in the interior of New South Wales and Queensland.

▲ *Ctenotus allotropis*. Bollon, Qld.
▶ *Ctenotus agrestis*. Aramac, Qld.

Ctenotus ariadnae has more stripes than most striped skinks. Its 18–20 narrow pale stripes are sharp and straight on its back, ill-defined on the flanks and tend to break up around the shoulders. This is a desert specialist, living among spinifex on sand flats.

▲ *Ctenotus ariadnae.* Hay River, NT.

▼ Copper-tailed Skink (*Ctenotus taeniolatus*). Sydney, NSW.

The **Copper-tailed Skinks** (*Ctenotus taeniolatus*) are boldly marked with simple stripes, though there is almost always a small pale spot behind the eye. They occupy a variety of habitats along the east coast. Those from rocky habitats dig shallow burrows under rocks. Lizards from around Sydney tend to have the brightest copper flush on their tails.

▲ Eastern Striped Skink (*Ctenotus robustus*). Brisbane, Qld.

The **Eastern Striped Skink** (*Ctenotus robustus*) is one of the most variable and widespread striped skinks. It penetrates the south-east further than any other, reaching the Melbourne area, and extends to the far north of Western Australia. Most have boldly striped backs and diffuse pale spots along the flanks. On some sandy islands off the east coast, completely patternless individuals live alongside normal striped lizards.

▲ Eastern Striped Skink (*Ctenotus robustus*). North Stradbroke Island, Qld.

▼ Lancelin Island Skink (*Ctenotus lancelini*). Lancelin, WA. B. Bush

The **Lancelin Island Skink** (*Ctenotus lancelini*) has a complex pattern of stripes, spots and dashes, and very restricted range. Until recently it was believed to be confined to a 9 hectare island off the west coast north of Perth. The lizard pictured was collected at nearby Lancelin. It is the first recorded mainland specimen.

▲ Leopard Skink (*Ctenotus pantherinus*). Halls Creek, WA.

◄ Carpentarian Ctenotus (*Ctenotus striaticeps*). Mt Isa area, Qld.

The **Carpentarian Ctenotus** (*Ctenotus striaticeps*) is arguably the most beautiful of all striped skinks. Its simple, orange-flushed pale stripes extend forward to the snout and contrast sharply against a black background. It lives in semi-arid woodlands on stony soil in the northern border area of Northern Territory and Queensland.

The **Leopard Skink** (*Ctenotus pantherinus*) is distinctively patterned with rows of pale-edged dark spots. There are several races which, between them, cover virtually the whole semi-arid to arid zone – about two-thirds of Australia.

Bands in the sand: desert sand-swimmers

▲ Broad-banded Sand-swimmer (*Eremiascincus richardsonii*). Port Germein, SA.

The two species of sand-swimmers (*Eremiascincus*) do not exhibit any of the features normally associated with burrowing skinks, yet they are named for their ability to dive into loose sand and vanish beneath the surface. They have well-developed limbs with a full complement of five digits, their eyes are large and ear-openings are exposed, their bodies are not elongated and their snouts are rounded rather than shovel-shaped. Their highly polished glossy scales may reduce friction, while a series of low, smooth longitudinal ridges along the rear back and tail in some populations might aid movement through sand. Both have a pattern of cross-bands – usually prominent, but if weak then some indication is normally visible on the tail.

Sand-swimmers are crepuscular to nocturnal and spend most of their active time foraging on the surface like typical skinks. By day they shelter beneath leaf litter or in burrows, including those of larger animals such as rabbits. They take a wide range of prey including smaller lizards, and seem able to use their long tails as a fat storage.

They live in arid regions across Australia, extending into moister well-drained timbered habitats at the eastern and western limits of their ranges. They are not restricted to soft sandy soils, and can be found on a wide variety of firm soil types. Both species are egg-layers.

The **Broad-banded Sand-swimmer** (*Eremiascincus richardsonii*) is prominently marked with 8–14 dark bands across its body and 19–32 across its original tail.

The **Narrow-banded Sand-swimmer** (*Eremiascincus fasciolatus*) is often more weakly patterned with narrower bands. It has 10–19 across the body and 35–40 across the original tail. It is common on desert sand dunes, where some nearly patternless populations are known as Ghost Skinks. When disturbed it readily dives into loose sand but if pursued beneath the surface it 'pops up', sprints and then dives back into the substrate.

▲ Broad-banded Sand-swimmer (*Eremiascincus richardsonii*). Ashford area, NSW. G. Swan.

▼ Narrow-banded Sand-swimmer (*Eremiascincus fasciolatus*). Bullara Station, WA.

Clipping the tip: Cape York Tree and Beach skinks and the Sheen Skink

▲ Cape York Beach Skink (*Emoia atrocostata*). Saibai Island, Qld.

In many respects Torres Strait and the far northern tip of Cape York have as much to do with New Guinea and the western Pacific as they do with Australia. Within that realm are a number of animals and plants that are not otherwise associated with this continent. As their broad ranges sweep across to our north, a few species just clip the tip. To survive here they must jockey for space among the plethora of indigenous species.

Beach skinks

Emoia are tropical skinks, with more than 100 species occupying an immense arc from the Pacific islands to Asia. In many areas across this vast tract they are the most numerous and diverse terrestrial and arboreal lizards.

The broad tropical distribution of *Emoia* just clips the tip of Cape York and the Torres Strait islands. Two species occur there, both of which have much wider ranges beyond Australia. They are moderately large, slender skinks with movable eyelids enclosing a transparent spectacle, well-developed limbs each with five long slender digits, slender whip-like tails and long pointed snouts. Both are egg-layers, with a fixed clutch size of two.

The **Cape York Beach Skink** (*Emoia atrocostata*) lives among beach rocks and mangroves. It clings so closely to the

coastline that a normal part of its foraging behaviour includes dodging the incoming waves and hunting invertebrates in their wake. Not surprisingly, it is a powerful swimmer and will not hesitate to take to water if pursued.

The **Cape York Tree Skink** (*Emoia longicauda*) is arboreal, foraging on slender branches and among foliage in forest clearings and margins. It is extremely agile and able to leap between branches, aided by specialised scales beneath its digits. These are extremely fine and numerous, and may bear microscopic filaments called setae, much like those on the padded toes of some geckos.

Sheen Skink

The genus *Eugongylus* also just manages to include Cape York in its broad tropical range from the Solomon Islands to eastern Indonesia. Just one wide-ranging species, the **Sheen Skink** (*Eugongylus rufescens*)

▲ Cape York Tree Skink (*Emoia longicauda*). Moa Island, Qld.

▼ Sheen Skink (*Eugongylus rufescens*). Lockerbie Scrub, Qld.

lives in closed forests and areas with thick low cover from the tip of the Cape, through Torres Strait to New Guinea. Adults, with a head and body length of 17 centimetres, are uniform brown with an iridescent sheen. In contrast, the young are brightly banded. Sheen Skinks live in rotting logs, among thick leaf litter, in tree hollows and disused burrows. Their long bodies and short limbs are ideal for wriggling through tight cavities. They are crepuscular to nocturnal egg-layers.

Stranded: castaways on mountains and islands

▲ Mount Bartle Frere Skink (*Techmarscincus jigurru*).
Mt Bartle Frere, Qld.

Isolation on mountain-tops and islands is
a mixed blessing. Separation from all
relatives reduces competition and offers
protection from some predators. It is also a
fine ingredient for the evolution of new
species, but there is no 'Plan B' should
unfavourable events occur.

The microclimate of a mountain-top can
be as effectively isolated by the surrounding
warmer lowlands as an island is by an
ocean. Some Australian reptile species have
restricted distributions encompassed within
cool upland enclaves. Their remote habitats
act as refuges for relict populations whose
ancestors, in past cooler times, enjoyed
broader distributions over more extensive
temperate landscapes.

Mount Bartle Frere Skink

Restricted to the lofty, mist-enshrouded
outcrops of bare rock, set among windblown
heaths and stunted moss forests at an
altitude of 1440–1620 metres on the
summit of Queensland's highest peak,
is the **Mount Bartle Frere Skink**
(*Techmarscincus jigurru*). Below it lies an
impenetrable barrier of warm tropical
lowland rainforest. The skink is the sole
member of its genus but its closest relatives
are probably more southerly groups such as
the Beech Skink (*Harrisoniascincus*).

In common with some other rock-dwelling
skinks that bask and forage on exposed
surfaces, it has a flattened body, long limbs
and digits. The flattened body is also ideal for
sheltering in narrow rock crevices. It is an egg-
layer, producing about four eggs per clutch.

Beech Skink

Above about 1000 metres in northern New South Wales and adjacent south-eastern Queensland, subtropical rainforest gives way to cool temperate forest communities including the Antarctic Beech (*Nothofagus moorei*). The **Beech Skink** (*Harrisoniascincus zia*) is confined to these areas. It is a sun-loving species, commonly seen basking among leaf litter beside walking tracks.

Though superficially similar to some of the garden and grass skinks (*Lampropholis*), the Beech Skink has a bright yellow belly. It is an egg-layer.

Lord Howe Island Skink

There can be no denying the effectiveness of oceans as barriers. Residents of islands can reach them three ways – their ancestors occurred there while the island was connected to a larger landmass, they were transported by humans, or they swam or rafted on floating debris.

Lord Howe and Norfolk Islands and their associated islets were formed in situ by volcanic activity, and the **Lord Howe Island Skink** (*Cyclodina lichenigera*) living there predates human activity in the region. The species may have evolved in situ but its ancestors must have reached these scattered islands of their own accord by clinging to debris. The skinks are now restricted to those islands but their closest relatives are probably other *Cyclodina* species living in New Zealand.

The skinks have suffered from human activity, particularly feral introductions and habitat degradation. They are now extinct on Norfolk Island and very rare on Lord Howe

▲ Beech Skink (*Harrisoniascincus zia*). Lamington Plateau, Qld.

▲ Lord Howe Island Skink (*Cyclodina lichenigera*). Blackburn Island.

Island, where they have been replaced by Garden Skinks (*Lampropholis delicata*) introduced from the Australian mainland. They range from rare to abundant on offshore rocks and islands, depending on levels of disturbance and the presence or absence of rats.

These skinks are crepuscular egg-layers. They probably eat a variety of arthropods, and there is a record of a lizard rolling a tern's egg until it broke, then devouring the contents.

That bug-eyed appeal: geckos

▲ Golden-tailed Gecko (*Strophurus taenicauda*). Lake Broadwater, Qld.

It is probably their large unblinking eyes, padded feet and soft velvety skin that create the special appeal unique to geckos. Those features strike a nurturing chord within us, combining to generate impressions of harmlessness, infancy and sensitivity.

All Australian geckos have clear spectacles covering their lidless eyes, and use their broad flat tongues like damp windscreen-wipers to keep them clean. They are among the few lizards able to vocalise, but for the most part, the squeaks and barks audible to us humans are reserved for moments of high stress – including social conflict, mating or attack by predators. The introduced Asian House Gecko is the most garrulous gecko in Australia, uttering its 'chuck … chuck … chuck' call day or night as a statement of occupancy.

Australian geckos are all egg-layers, and the clutch is almost always fixed at two, occasionally one. A worldwide family of geckos called the Gekkonids, which includes many Australian species, lay eggs with hard, brittle shells. However, more than two-thirds of Australian geckos belong to a regionally endemic group called the Diplodactylids and lay soft, parchment-shelled eggs like most other lizards.

Included within both groups are geckos with simple clawed feet and others with padded toes. Beneath the pads are rows of broad plate-like scales, each covered by thousands of microscopic bristles with branched, flattened tips. These structures called setae provide an enormous surface

▲ Prickly Knob-tailed Gecko (*Nephrurus asper*). Collinsville area, Qld.

▼ Eyed Velvet Gecko (*Oedura monilis*). Chesterton Range, Qld.

area of contact and are believed to adhere at the molecular level. This means the gecko shinning across your ceiling is not clinging by suction but by an intimate bond with the substrate. Its sure grip and ability to rapidly engage and disengage are the focus of scientific research into the quest for dry adhesives.

Geckos have proven themselves a robust and versatile group, able to exploit diverse and hostile environments. In the shimmering midday summer heat, nothing moves on a desert sand dune. With surface temperatures exceeding 60° Celsius, exposure would mean swift and certain death. But when the sun sets, geckos emerge to patrol the cooling sand. Big-eyed, translucent-skinned and thin-limbed, it seems inconceivable that such fragile-looking animals could have devised means of existing in such a place. Their skill lies in avoiding rather than tolerating the extremes.

Geckos thrive on windswept south coastal granites, cast away on far-flung islands, and in the darkening gloom of tropical rainforests. They hide in cavities behind loose bark and in rock crevices, they commandeer disused insect and spider holes and cling to the foliage of shrubs and tussocks. They occur across the continent except parts of the south-east and Tasmania. Geckos are most diverse in warm dry areas, particularly tropical woodlands, spinifex deserts, rock outcrops and escarpments.

Meanwhile on the urban scene, while customers queue for their orders in bustling inner city takeaways, the overhead fluorescent lights attract an assortment of moths and crickets. They make fine hunting grounds for house geckos that stalk their quarry with gravity-defying ease across the ceilings and walls.

Geckos may well be soft-skinned and appear fragile, yet with 120 named Australian species, they are second only to skinks in terms of diversity. The family includes many discreet, specialised and poorly known species, but their success within the human environment is without parallel. They use our buildings for shelter, dine on the insects we attract, and employ our transport systems as a global dispersal service. It is worth remembering that the world's most widespread and invasive lizard is a gecko.

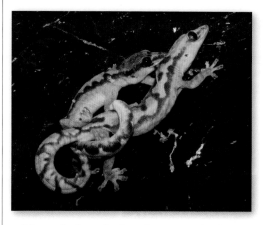

▲ It is standard practice for males to grasp females by the nape when mating. Velvet Geckos (*Oedura robusta*). Kurwongbah, Qld.

▼ Rough-throated Leaf-tail (*Saltuarius salebrosus*). Cracow, Qld.

Camouflage artists: leaf-tailed geckos

▲ Rough-throated Leaf-tail (*Saltuarius salebrosus*). Blackdown Tableland, Qld.

The 16 species of leaf-tailed geckos turn camouflage into an art form. At night they rest motionless on tree trunks and rock faces. Their flecked or marbled patterns blend perfectly with their variegated backgrounds, while their flat heads and bodies, rough textures and wide, flat tails render them virtually invisible against the substrate.

These impressive geckos live in rainforests, vine thickets, and major outcrops of sandstone and granite in eastern Australia between Sydney and Cape York. They are sedentary lizards that shelter in narrow crevices, tree hollows, and the walls and ceilings of caves.

Leaf-tailed geckos are usually seen clinging to rough vertical surfaces. They lack adhesive pads but their long slender limbs and angular, clawed, bird-like digits ensure a secure grip. There is normally a high proportion of lizards with regenerated tails but, given their sedentary habits and superb camouflage, most of these are more likely to have resulted from disputes with other individuals than attacks by would-be predators. Original tails are spiny with pointed tips while regenerated tails are smooth with round tips.

The largest and most flamboyant of the leaf-tails are the six species of *Saltuarius*, with generally complex marbled patterns and elaborately flanged tails.They range from northern New South Wales to the Wet Tropics of north-eastern Queensland.

▲ Granite Belt Leaf-tail (*Saltuarius wyberba*). Ballendean, Qld.

◄ Broad-tailed Gecko (*Phyllurus platurus*). Lane Cove NP, NSW.

The **Rough-throated Leaf-tail** (*Saltuarius salebrosus*) is the most widespread leaf-tail, occurring on sandstone cliffs and caves in central-eastern Queensland. The lizard pictured on page 83, with a head and body length of 14 centimetres, shows its true size against an ochre hand stencil in a sheltered gallery of Aboriginal rock art. The smoothly rounded tail is regenerated. The lizard on page 82 has a pointed original tail.

The **Granite Belt Leaf-tail** (*Saltuarius wyberba*) perfectly matches the irregular flecked surfaces of granite boulders. It lives in an area of upland granite outcrops on the New England Tableland of New South Wales and adjacent Queensland. This lizard has an regenerated tail with a rounded tip.

The nine species of *Phyllurus* are characterised by flecked patterns and simpler tails. They extend from north-eastern Queensland to Sydney. The **Broad-tailed Gecko** (*Phyllurus platurus*) thrives in the sandstone cliffs, caves and overhangs around Sydney and even resides on house walls, in garages and basements when these abut suitable habitat. Large numbers sometimes cohabit, and some caves are festooned with their shed skins.

In tropical mid-eastern Queensland a cluster of leaf-tailed gecko species in the genus *Phyllurus* is confined to narrow distributions, each in a rainforest separated from its relatives by intervening dry terrain. **Phyllurus isis** is known only from dense low rainforest on the rocky summit of Mt Blackwood.

The **Ringed Thin-tailed Gecko** (*Phyllurus caudiannulatus*) is one of a small group of leaf-tailed geckos with narrow cylindrical tails. It is restricted to dense rainforest in the vicinity of Bulburin State Forest in mid-eastern Queensland. This arboreal gecko favours the complex cavities within the trunks of mature fig trees. A suitable tree may harbour a large colony. The two lizards pictured below, exhibiting original and regenerated tails, were among a group of 13 individuals counted on one tree.

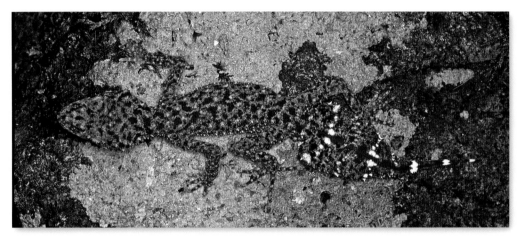

▲ *Phyllurus isis*. Mt Blackwood, Qld.

▼ Ringed Thin-tailed Gecko (*Phyllurus caudiannulatus*). Bulburin State Forest, Qld.

The **Cape York Leaf-tail** (*Orraya occulta*) is an unusual species and the sole member of its genus. This is the most northerly leaf-tail, isolated in the McIlwraith Range near Coen on Cape York. Only a handful of specimens have ever been found, due partly to its remote location and also because of an apparent scarcity within the known habitat. It is associated with large boulders under rainforest along a creek.

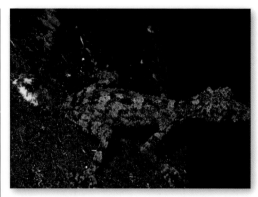

▲ Cape York Leaf-tail (*Orraya occulta*). Peach Creek, McIlwraith Range, Qld.

▲ Cape York Leaf-tail (*Orraya occulta*). Peach Creek, McIlwraith Range, Qld.

Barking mad: thick-tailed and knob-tailed geckos

▲ Common Thick-tailed Gecko (*Underwoodisaurus milii*) with original tail. Helidon, Qld.

All geckos have voices, but few employ them with such startling effect as the thick-tailed and knob-tailed geckos. When provoked, these large-headed lizards raise and sway their plump bodies high on slender limbs, sinuously wave their tails and suddenly leap at their aggressor with mouth agape, uttering a loud rasping bark. This has earned them the name 'barking geckos'. To the human observer this gruff display looks comical and even endearing, but small predators like dunnarts would no doubt pause to consider their options.

These terrestrial geckos lack any adhesive pads on their short thick digits. They shelter by day under rocks and in burrows, emerging at night to hunt insects, spiders, scorpions and even other geckos.

Two species of thick-tailed geckos (*Underwoodisaurus*) have plump, carrot-shaped original tails with pointed tips and sharp-edged, pale bands, while the nine species of knob-tailed geckos (*Nephrurus*) have a curious spherical enlargement on the tail-tip. The purpose of this distinctive fleshy 'knob', one of the most unusual appendages of any reptile, remains an ongoing mystery. All regenerated tails are swollen and blunt with no pointed tips or knobs, and little or no pattern.

Thick-tailed geckos live mainly in dry to well-drained habitats across southern

▲ Common Thick-tailed Gecko (*Underwoodisaurus milii*) with regenerated tail. Mt Dale, WA.

Australia, and knob-tailed geckos occupy the deserts and northern tropical woodlands and outcrops. There is overlap between the two groups in some areas, but they largely replace each other geographically.

Thick-tailed geckos

The **Common Thick-tailed Gecko**
(*Underwoodisaurus milii*) spans southern Australia in a range of habitats from coastal heath to granite and sandstone outcrops. At some localities it so abundant that virtually every suitable rock harbours one or more resident lizards. Thirteen were recorded under a single granite slab on an outcrop in central Victoria. The differences between original and regenerated tails on the lizards pictured above and on page 87 are obvious.

The **Border Thick-tailed Gecko**
(*Underwoodisaurus sphyrurus*) is not as easy to find. It has a much more restricted distribution on the New England Tablelands in northern New South Wales and adjacent Queensland. It is mainly associated with

granite outcrops in an area that experiences mild to hot summers and numbing winter frosts with occasional snowfalls. The individual here clearly demonstrates what the group does best – rearing, lunging, barking and looking fierce.

▼ Border Thick-tailed Gecko (*Underwoodisaurus sphyrurus*). Ballendean, Qld.

Knob-tailed geckos

The minuscule tail of this **Prickly Knob-tailed Gecko** (*Nephrurus asper*) is not regenerated. It is original, but reduced to little more than a bump on its rump, complete with a spherical knob. This is one of three similar species that stand alone among all Australian (and possibly world) geckos in being unable to discard their tails and grow new ones. It appears that evolution has already done the job, leaving this bizarre gecko with little to lose. Prickly Knob-tailed Geckos live in dry woodlands and outcrops of Queensland.

The **Smooth Knob-tailed Gecko** (*Nephrurus levis*) has a fleshy tail that serves as a fat storage. Three separate races occupy the deserts from inland New South Wales to the west coast. All live mainly on sandy spinifex plains, sheltering by day in burrows of their own construction or those abandoned by other animals. The disused burrows of Central Netted Dragons (*Ctenophorus nuchalis*) are particular favourites. *Nephrurus levis pilbarensis* is confined to the north-west. The specimen pictured on page 90 is using its fleshy tongue to clean the transparent spectacles covering its eye. *N. levis levis* is the most widespread form, covering most of the species' range.

▼ Prickly Knob-tailed Gecko (*Nephrurus asper*). Moranbah, Qld.

▲ A subspecies of the Smooth Knob-tailed Gecko, *Nephrurus levis pilbarensis* occurs in the Western Australian Pilbara. De Grey River, WA.

▼ *Nephrurus levis levis* occurs over vast tracts of the interior. Currawinya NP, Qld.

Forest wraith: the Chameleon Gecko

▲ Chameleon Gecko (*Carphodactylus laevis*). Viewed front-on, the Chameleon Gecko looks wraith-like, or even 'extra-terrestrial'. Lake Barrine, Qld.

The spectacular **Chameleon Gecko** (*Carphodactylus laevis*) is a secretive inhabitant of north Queensland's tropical rainforests. It probably earns its common name from its distinctive, laterally flattened body with an acute vertebral ridge. It also has slow deliberate movements like its namesake.

This lizard, with a head and body length of 13 centimetres, hides by day under leaf litter and in hollow limbs and cavities within tree buttresses. It is probably mainly terrestrial, stalking invertebrates at night on the forest floor. Yet it frequently climbs slender saplings, taking up a position head downwards about a half a metre from the ground. This offers an excellent opportunity to scan with its large, forward-directed eyes for anything moving below.

The Chameleon Gecko is related to the leaf-tailed, thick-tailed and knob-tailed geckos. When members of this group lose their tails it is always broken at the base. Many geckos can break half of the tail, or forfeit only the tip if necessary, but for this group tail-loss is a case of all or nothing.

The Chameleon Gecko's original tail is carrot-shaped, slightly laterally compressed and black with sharp white bands. The regenerated tail is simpler with a mottled pattern. The tail is extremely fragile and easily discarded. Like all lizards' tails it takes on a life of its own when severed, wriggling and twisting while the lizard makes good its escape. But when the regenerated tail of a Chameleon Gecko is broken it has another unique trick. It emits a loud squeaking noise as it moves. It is not known whether original tails make the same noise when they are discarded, but no other lizards are recorded to have noisy broken tails.

▲ Chameleon Gecko (*Carphodactylus laevis*). The clear white bands indicate this gecko has an original tail. Lake Barrine, Qld.

▶ Chameleon Gecko. This gecko has adopted an ambush posture, head downwards on a sapling. The mottled tail is regenerated. Wooroonooran NP, Qld.

An agile giant: the Ring-tailed Gecko

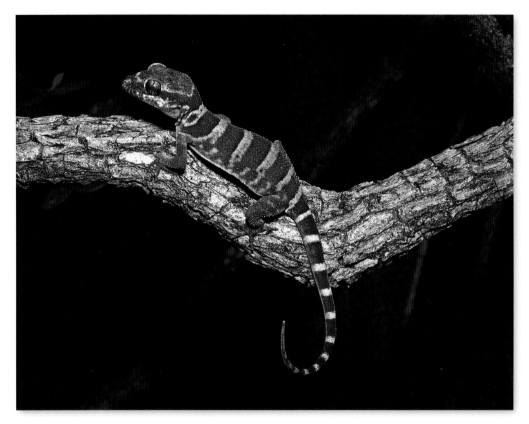

▲ Ring-tailed Gecko (*Cyrtodactylus louisiadensis*). Chillagoe, Qld.

With a head and body length of 16 centimetres and a total length of more than 30 centimetres, the **Ring-tailed Gecko** (*Cyrtodactylus louisiadensis*) is an impressive lizard by any reckoning. It is one of Australia's largest geckos, and a voracious predator that includes other geckos and even frogs in its diet.

The Ring-tailed Gecko is extremely swift and agile, climbing with ease over coarse rock faces and among vines. Its padless digits are slender, angular and well-clawed, while the long thin tail can be employed as a counterweight to aid balance, enabling it to make well-coordinated leaps.

The Ring-tailed Gecko is the only Australian member of its genus, but many close relatives are common in rainforests from Asia to the western Pacific. The Australian species lives in north Queensland, New Guinea and the Solomon Islands. It is mainly rock-inhabiting, extending north from the exposed boulder slopes and limestone karsts of the Chillagoe district to the rainforest-lined rocky watercourses of the McIlwraith Range in north-eastern Cape York. It also occupies buildings where these abut suitable habitat. This species belongs to a worldwide group of geckos that lay a pair of round, brittle-shelled eggs.

Sticky business: the pad-tailed geckos

Few animals are designed to stick as effectively as the pad-tailed geckos *(Pseudothecadactylus)* of northern Australia. These large impressive geckos, with head and body lengths up to 12 centimetres, have each digit flattened along its whole length to form a very large pad, equipped below with a series of broad transverse plates from base to tip. The slender, cylindrical tail is prehensile and tipped with another pad. Large scales on the tail-tip probably feature the same type of microscopic bristles that provide adhesion to the digital pads of other geckos. Small wonder the three species are able climbers.

The **Giant Tree Gecko** (*Pseudothecadactylus australis*) from northern Cape York and the southern islands of Torres Strait shelters by day in tree hollows. The lizards have relatively deep heads and bodies and they prefer tight entrances that are just sufficient

▲▼ Adhesive structures beneath the digits and tail-tip ensure a secure grip when they climb.

to squeeze through. This probably helps protect them from predators like Spotted Tree Monitors (*Varanus scalaris*; see page 165) that spend much of their time investigating cracks and cavities. If disturbed, or even when an occupied hollow is approached, the geckos often betray their presence with a gruff bark.

▲ Giant Tree Gecko (*Pseudothecadactylus australis*). Prince of Wales Island, Qld.

▼ Giant Cave Gecko (*Pseudothecadactylus lindneri*). Arnhem Escarpment, NT.

The escarpments and rock art galleries of Arnhem Land are home to the **Giant Cave Gecko** (*Pseudothecadactylus lindneri*). At night they emerge from caves and crevices to hunt on the rock faces, and forage among the branches and foliage of adjacent trees. In addition to invertebrates, they also take nectar and lick oozing sap.

Lizard from Lilliput: the Clawless Gecko

▲ *Crenadactylus ocellatus ocellatus.* Upper Swan, WA.
B. Maryan

▼ *Crenadactylus ocellatus rostralis.* Kununurra, WA.

The tiniest gecko in Australia is the
Clawless Gecko (*Crenadactylus ocellatus*).
As the name indicates, this is the only
Australian gecko without claws. In their
place, the digits are each equipped with a
pair of large circular pads.

The four named subspecies of this
widespread, variable species have head and
body lengths ranging from 3 to 3.5
centimetres. They extend from the mallee
woodlands and shrublands of the south-west
to the arid central ranges and the sandstone
plateaus in the north of the Northern
Territory and Western Australia. Populations
differ so significantly in appearance and
behaviour that they probably represent
separate species. They are rarely
encountered active at night, and it is likely
that much of their feeding occurs within
their shelter sites. They lay pairs of
parchment-shelled eggs.

The south-western subspecies,
Crenadactylus ocellatus ocellatus, occupies a
very wide range of habitats, including many
offshore islands. This is a mainly terrestrial
gecko, sheltering under logs, leaf litter and
any available debris. It is patterned with
spots and streaks.

Other populations have uniformly striped
patterns. They live in and under clumps of
spinifex, climbing through the lattice of
needle-sharp spines and among the humid
cavities within their bases. *Crenadactylus
ocellatus rostralis* is associated with
sandstone outcrops in the southern
Kimberley region of Western Australia and
the adjacent Northern Territory.

Prickly customers: Bynoe's and nactus geckos

▲ Bynoe's Gecko (*Heteronotia binoei*). Westmar, Qld.

Bynoe's Gecko (*Heteronotia binoei*) is probably the most abundant lizard in dry habitats throughout Australia. These small, swift ground geckos have slender padless digits and a rough skin texture composed of small granular scales mixed with larger, pyramid-shaped scales. They have fared extremely well from European settlement and are particularly fond of old rubbish heaps. Throughout the inland, turn any discarded items and there is a strong chance that, even if nothing else is present, there will at least be a Bynoe's Gecko hiding underneath.

While there are currently three named species of *Heteronotia*, genetic studies reveal that Bynoe's Gecko is not just one widespread species but a complex of distinct forms including some all-female populations. The ranges of these parthenogenetic populations overlap with several different lineages of more typical bisexual lizards. It is believed the parthenogenetic geckos arose through hybridisation between genetically distinct populations. These common little geckos are a complicated mix of species with no

◄ A banded colour form of Bynoe's Gecko (*Heteronotia binoei*). Lawn Hill, Qld.

▼ Cape York Nactus (*Nactus eboracensis*). Tip of Cape York, Qld.

published information to clearly separate and describe them.

In north-eastern Queensland and the Torres Strait islands, the four species of nactus geckos (*Nactus*) are common beneath ground debris. They are very similar to Bynoe's Geckos and sometimes live alongside them. The enlarged scales on their backs are conical rather than pyramid-shaped. The **Cape York Nactus** (*Nactus eboracensis*) is the most abundant ground gecko north of Princess Charlotte Bay, extending to some southern Torres Strait islands.

With a slender body, long thin limbs and enormous eyes, the **Black Mountain Gecko** (*Nactus galgajuga*) is an agile and extremely swift rock-hopper. It lives only among the piled granite boulders of Black Mountain near Cooktown. The gecko needs all the dexterity it can muster, for it shares the unique habitat with a voracious predator, the Ring-tailed Gecko (*Cyrtodactylus louisiadensis*), that readily devours other geckos when it can catch them.

▲ Black Mountain Gecko (*Nactus galgajuga*). Black Mountain, Qld.

Alien invaders: Asian House and Mourning geckos

▲ The ultimate urbanite! An Asian House Gecko (*Hemidactylus frenatus*). Kurwongbah, Qld.

Employing us as dispersal agents, the **Asian House Gecko** (*Hemidactylus frenatus*) has become the world's most widespread lizard. It enjoys a pantropical distribution, founded by stowaways among cargo.

Over much of its vast range it lives exclusively alongside humans, patrolling our walls and ceilings, hiding behind our furnishings, and stationing itself beside outdoor lights to snatch insects as they arrive. Cavities within our walls are repositories for its eggs. The structure of those round eggs contributes to its dispersal success. The hard shells are more resistant to dehydration than the soft parchment-like shells of most lizard eggs and are well suited to survive voyages.

We do not know when Asian House Geckos first arrived from South-east Asia, but they have been residing in Darwin since at least the 1960s. For many years they were restricted to that region. By the 1980s individual waifs started turning up at the Port of Brisbane. These were probably pioneers of a separate invasion, delivered among shipping containers, rather than migrants from the established Darwin population.

Numbers remained low until the mid-1990s when they dispersed rapidly across the city and to towns north, west and south. They are very vocal geckos, and their distinctive 'chuck … chuck … chuck' call has become a normal and accepted background noise in virtually every dwelling along the Queensland coast, and is being heard with increasing frequency as far south as Taree, New South Wales and west through the Kimberley.

It is difficult to reconcile this invasive reptile as benign, given its reputation for displacing native species. It has replaced the Northern Dtella (*Gehyra australis*) as the common house gecko in Darwin, and is applying pressure to the Dubious Dtella (*Gehyra dubia*) on dwellings in Townsville. There is also disturbing evidence from the Northern Territory of populations becoming established in bushland away from human dwellings, offering these invasive geckos greater opportunities to displace native species.

The **Mourning Gecko** (*Lepidodactylus lugubris*) is probably also a migrant to our shores, and populations are widespread through island nations of the western Pacific. In Australia it occurs in north-eastern Queensland, has recently spread to some resort islands on the Great Barrier Reef, and is now turning up in Darwin.

This is a parthenogenetic species, with enormous advantages for dispersal, as a single egg may be all that is needed to establish a new population. Being much smaller than the more aggressive Asian House Gecko, it is more discreet and secretive where the two occur together.

▲ The Asian House Geckos' (*Hemidactylus frenatus*) round, hard-shelled eggs are resistant to dehydration and ideally suited for transportation. Kurwongbah, Qld.

▼ The Asian House Geckos' iris has distinctly scalloped margins. Kurwongbah, Qld.

▲ Mourning Gecko (*Lepidodactylus lugubris*). Cairns, Qld.

Trees and rocks to houses: dtellas and velvet geckos

▲ Common Dtella (*Gehyra variegata*). Currawinya NP, Qld.

Well-padded digits and relatively flat bodies are an excellent combination for geckos that shelter in narrow vertical gaps on trees and rocks. Velvet geckos (*Oedura*) and dtellas (*Gehyra*) live under the bark of dead trees and in rock crevices across most of Australia. Both groups include tree- and rock-dwelling specialists, but most species happily occupy both types of shelter. Some often enter our homes, where gaps behind bookcases and wardrobes offer cover and lights attract insect prey. They can move easily over smooth vertical surfaces, including windows.

Dtellas

Many of the 17 species of *Gehyra* are outwardly similar, and a few pose real identification problems for experts. They are generally grey to orange with prominent pale spots or a network of dark lines, and they have large, flat, semi-circular pads with a claw arising from the upper surface on all but the clawless inner digits. Most are moderate sized geckos with head and body lengths of about 6 centimetres.

The **Common Dtella** (*Gehyra variegata*) is one of the most abundant vertebrates across its vast range between the eastern interior of Queensland and the west coast. It lives in dry habitats, where it is as much at home under any debris lying on the ground as it is on rocks and trees. It is a common house gecko throughout the inland.

▲ Dubious Dtella (*Gehyra dubia*). Brisbane, Qld.

Small wonder geckos are so at home in our houses. The **Dubious Dtella** (*Gehyra dubia*) under the watchful eye of young Amy Couper is as comfortable on a glass window as it is on a wall, rock or tree trunk. This is a common eastern Australian house gecko, from central New South Wales to north Queensland. Its favourite natural habitat is dry forests, woodlands and outcrops.

Velvet geckos

The 15 species of velvet geckos (*Oedura*) are generally easier to identify, as each tends to have its own distinctive pattern. Most are marked with either small pale spots, a row of large pale blotches, or prominent bands. They include moderate to large species, with head and body lengths ranging from 6 centimetres to more than 10 centimetres. Velvet geckos' tails vary from narrow and cylindrical to very fleshy and slightly depressed. Some species probably use them as a fat storage. The group is named because of the uniformly granular, soft velvety skin.

▲ The Spotted Velvet Gecko (*Oedura tryoni*) lives on trees and rocks from the tablelands of northern New South Wales to the dry woodlands in the interior of southern Queensland. Mt Nebo, Qld.

▲ Northern Velvet Gecko (*Oedura castelnaui*) juvenile. Collinsville, Qld.

The **Northern Velvet Gecko** (*Oedura castelnaui*) lives behind the loose bark of dead trees in woodlands of tropical north-eastern Queensland. It has also been found sheltering behind the dead foliage skirting grass trees. The two lizards pictured here look quite different, as the prominent bands of juveniles are composed of clean, unbroken colour. As they grow, some elements of the pattern begin to fragment. This is a common developmental pattern on some of the larger velvet gecko species.

▲ Northern Velvet Gecko (*Oedura castelnaui*) adult. White Mountains NP, Qld.

The **Clouded Gecko** (*Oedura jacovae*) is a newly named species. It was only described in late 2007, although its distribution is centred largely on the Brisbane area where it has been well known for many years. It is arboreal, favouring woodlands that include ironbarks. In some areas it lives as a house gecko, but only where homes are adjacent to bushland.

Living on house walls has clear benefits in the form of insects attracted to lights. The **Robust Velvet Gecko** (*Oedura robusta*) pictured has captured a cicada.

The **Fringe-toed Velvet Gecko** (*Oedura felicipoda*) has a prominent fringe of scales edging each digit. It is an exclusively rock-inhabiting species, living in sandstone caves and cliffs of the northern Kimberley region, Western Australia.

▲ Clouded Gecko (*Oedura jacovae*). Southbrook, Qld.

▼ Robust Velvet Gecko (*Oedura robusta*). Kurwongbah, Qld.

► Fringe-toed Velvet Gecko (*Oedura felicipoda*). Mitchell Falls, WA.

Cold comfort: marbled geckos

▲ Common Marbled Gecko (*Christinus marmoratus*). Kangaroo Island, SA.

Geckos thrive best in arid and tropical regions but the marbled geckos (*Christinus*) occupy cool temperate habitats where most other geckos cannot live. They are the only geckos that occur in some of the most southerly parts of their range.

They live in narrow crevices under rocks, surface debris and behind loose bark across southern Australia, including many offshore islands. Marbled geckos lay a pair of round, brittle-shelled eggs. These are often deposited communally.

The three species share soft velvety skin and padded digits, each with a single line of broad plates under each digit, and a large circular pair under the tip of the digit. Their tails are long and fleshy, and may comprise nearly one-quarter of the total body mass. This represents a substantial repository for nutrients, and the complete loss of a tail would constitute a significant setback for an individual.

By far the most abundant and widespread species, the **Common Marbled Gecko** (*Christinus marmoratus*) covers much of the southern mainland. Urban populations are found in Perth, Adelaide and parts of Melbourne, hiding under boards and building materials. In many areas they are extremely common, with lizards occupying most suitable shelter sites. A common feature of juveniles is a row of orange blotches along the tail. These fade on adults.

The **Lord Howe Island Gecko** (*Christinus guentheri*) is not faring as well as its mainland counterparts. Isolated on Lord Howe Island and its satellites, and on islets off Norfolk Island, it has suffered serious declines from habitat degradation and the introduction of feral animals. Fortunately it remains common on less disturbed islets. In addition to insects, nectar is an important seasonal food source.

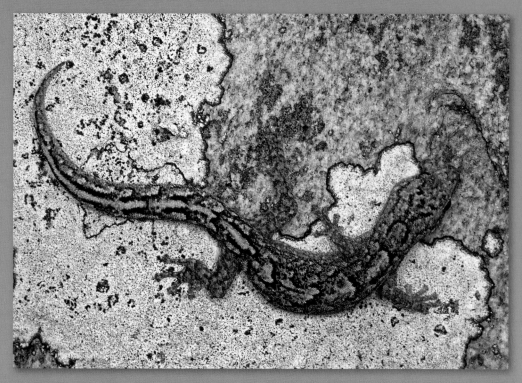

▲ Common Marbled Gecko (*Christinus marmoratus*). Rabbit Island, WA.

▼ Lord Howe Island Gecko (*Christinus guentheri*). Blackburn Island.

Feet firmly on the ground: ground, swift and Beaked geckos

▲ Fat-tailed Gecko (*Diplodactylus conspicillatus*). Moranbah, Qld.

These terrestrial geckos are mostly found on dry open terrain. They occur throughout mainland Australia except in the lower south-east and far northern Cape York. Most species are opportunist predators of small invertebrates but some appear to be specialist termite feeders. Many are also particular about where they live, sheltering by day in the vertical shafts of insect and spider holes. They belong to a group of geckos endemic to the Australian region that lay pairs of soft, parchment-shelled eggs.

Ground geckos

All 13 species of ground geckos (*Diplodactylus*) have padded digits, relatively short limbs and short fleshy tails. Though patterns vary greatly, there is a common trend for large pale dorsal blotches or a pale vertebral stripe. Most are not particularly swift, and many species adopt nightly ambush positions a few centimetres above the ground on fallen twigs. From these low vantage points they can pounce on passing insects.

Many species hide by day under small surface objects, and the presence of a gecko together with shed skins of different ages indicates long-term use of suitable shelter sites.

The **Fat-tailed Gecko** (*Diplodactylus conspicillatus*) is an unusual lizard with whorls of large squarish scales armouring its short flat tail. It is a specialist on several counts. It shelters exclusively in the vertical shafts of invertebrate holes, positioning itself head downwards and blocking the passage with its tough tail. It also eats only termites. Despite these peculiar features it has the broadest range of the ground geckos, covering about two-thirds of arid to dry tropical Australia in habitats ranging from featureless desert to tropical grassy woodlands.

The **Stone Gecko** (*Diplodactylus vittatus*) is a variable species with dorsal blotches, or a straight-edged to zigzagging vertebral stripe. That notched stripe probably reminded settlers of the venomous British Adder (*Vipera berus*), for an old colloquial name applied to this gecko is 'Wood Adder'. It covers a large area of south-eastern Australia, penetrating temperate regions more effectively than other ground geckos.

▲ Stone Gecko (*Diplodactylus vittatus*). Idalia NP, Qld.

The **Pretty Gecko** (*Diplodactylus pulcher*) occurs in a broad expanse of arid to semi-arid lands in southern Western Australia and adjacent South Australia. Two main colour forms occur; the blotched form pictured and a striped form with a single broad pale vertebral stripe. Both forms normally occur together. It feeds on termites.

▲ Pretty Gecko (*Diplodactylus pulcher*). Warburton, WA.

▲ Beaded Gecko (*Lucasium damaeum*). Queen Victoria Spring, WA. B. Maryan

◀ Box-patterned Gecko (*Lucasium steindachneri*). White Mountains NP, Qld.

Swift geckos

The ten species of swift geckos (*Lucasium*) are generally very alert lizards, quick to flee when illuminated by torchlight. They are slender with long limbs, relatively thin tails and slender digits, with or without a pair of enlarged plates under the tips. Swift geckos tend to forage in more open areas than their relatives the ground geckos. They are more likely to occupy vertical insect- and spider-holes and are only infrequently found hiding beneath surface debris.

The **Box-patterned Gecko** (*Lucasium steindachneri*) is common across the northern interior of New South Wales and inland Queensland, extending north to Coen. It is very general in habits, sheltering under surface debris and in burrows, and feeding on a variety of invertebrates. The gecko here is emerging from a spider burrow.

Beaded Geckos (*Lucasium damaeum*) are common in semi-arid southern Australia, particularly in sandy areas with spinifex. They live exclusively in burrows by day, including those dug by other lizards such as Painted Dragons. Combating males have been observed to approach each other with bodies held high, uttering a continuous chirping sound.

Beaked Gecko

The **Beaked Gecko** (*Rhynchoedura ornata*) is a distinctive species and sole member of its genus. It superficially resembles the swift geckos, though it tends to be longer bodied and shorter limbed, with the hallmark short, beak-like snout. It lives almost exclusively in burrows by day, favouring vertical insect and spider shafts. It is a termite specialist with a wide distribution across arid Australia.

▲ Beaked Gecko (*Rhynchoedura ornata*). Kings Creek Station, NT.

▼ A Beaked Gecko emerging from a spider burrow. Currawinya NP, Qld.

Goo-squirters: spiny-tailed, striped and Jewelled geckos

▲ Golden-tailed Gecko (*Strophurus taenicauda*). Lake Broadwater, Qld.

The 17 species of *Strophurus* are found over most of northern and central Australia, extending south to the semi-arid mallee woodlands. They favour dry habitats, shunning rainforests and other moist areas. All have padded digits that are mainly adapted to grip slender branches and foliage.

When facing extreme harassment these unusual geckos can emit sticky goo from pores along the back and upper surface of the tail. This defensive strategy is largely unknown in the reptile world. While some species dribble and smear the repellent liquid, others violently eject a stream of droplets half a metre or more.

The substance has the consistency of treacle when first emitted, drying to cobweb-like filaments after exposure to air. It is known to cause intense irritation if it contacts eyes, and may also prove effective in gumming up the mandibles of invertebrate predators such as centipedes. Curiously, most geckos are generally reluctant to eject the liquid unless under great duress. It may be metabolically costly to produce.

There are three main types of squirting geckos. The striped geckos have simple linear patterns while the Jewelled Gecko has distinctive spots. These species live only within clumps of spinifex or sedge, usually in arid areas. Spiny-tailed geckos and their allies generally hide within the foliage of trees and bushes, or behind loose bark. Some have enlarged tubercles or spines on their tails and most have intricately patterned irises with brightly-coloured rims.

▲ Northern Spiny-tailed Gecko (*Strophurus ciliaris aberrans*). North West Cape, WA.

While primarily nocturnal, many of these geckos are often encountered basking, sometimes in conditions so hot they would seem lethal.

With a head and body length of nearly 9 centimetres, the **Northern Spiny-tailed Gecko** (*Strophurus ciliaris*) is the largest of the goo-squirters. It has long spines along its tail and large spines above each eye. It lives mainly in woodlands across northern and inland Australia.

One of Australia's most beautiful lizards, the **Golden-tailed Gecko** (*Strophurus taenicauda*) lives in dry forests of eucalypt and native pine in the southern interior of Queensland. It is unmistakable, with its black and white latticed pattern, orange blaze along its tail and eyes like a pair of bright rubies.

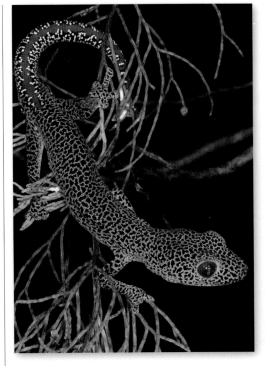

▲ Golden-tailed Gecko (*Strophurus taenicauda*). Lake Broadwater, Qld.

▲ Western Spiny-tailed Gecko (*Strophurus spinigerus*). Mosman Park, WA.

The **Western Spiny-tailed Gecko** (*Strophurus spinigerus*) is common in coastal heaths around Perth. A careful look in the garden shrubbery may reveal one clinging to a stem in filtered sunlight. It varies geographically, with yellow-eyed, orange-eyed and white-eyed forms occurring between the south-west and the mid-west coast.

The **Robust Striped Tail-squirter** (*Strophurus michaelseni*) lives in sedges and spinifex on sand plains along the west coast. The striped pattern provides ideal camouflage among the dense linear foliage

The **Jewelled Gecko** (*Strophurus elderi*) is found across the arid regions wherever spinifex grows. They live exclusively within its latticed foliage, seldom venturing beyond the protection of its sharp spines.

▲ Robust Striped Tail-squirter (*Strophurus michaelseni*). Marchagee, WA.

◀ Jewelled Gecko (*Strophurus elderi*). Paynes Find, WA.

The flap-footed lizards: just long, thin geckos?

▲ Burton's Snake-lizard (*Lialis burtonis*). Like their cousins the geckos, flap-footed lizards can lick their faces and lidless eyes with their tongues. Katherine area, NT.

At first glance no one could seriously treat these slender, apparently limbless snake-like reptiles as cousins of the soft-skinned, pad-toed geckos. Yet the two groups share a number of features that are so unique some scientists even classify them together in one family.

Like geckos, the flap-footed lizards employ a flat fleshy tongue to wipe clean the clear spectacle covering their lidless eyes. When harassed, handled or in some social contexts they can vocalise, uttering a loud squeaking sound. And they have a fixed clutch size of two eggs.

Flap-footed lizards are so named because the hind limbs have been reduced to simple scaly flaps. In some species these are barely obvious and scarcely larger than adjacent scales, but in others they are large and can be conspicuously moved. Forelimbs have been lost without trace.

They superficially resemble snakes, particularly when wriggling through vegetation. However, if disturbed on open ground, flap-footed lizards often move in a series of frenzied leaps, rather than with the sinuous sideways movements of snakes. Most also have distinct ear-openings (snakes have none), two rows of scales along the belly (most snakes have a single row), and tails as long as the body or up to four times longer (snakes have much shorter tails). Like many lizards they can discard and re-grow their tails. In extreme cases this could mean the loss of over two-thirds of an animal's total mass.

▲ Unlike snakes, most flap-footed lizards like this Helmeted Delma (*Delma mitella*) have clearly visible ear openings. Tolga, Qld.

▼ Common Scaly-foot (*Pygopus lepidopodus*). Big Desert, Vic.

Flap-footed lizards probably lost their limbs and developed long slender bodies to assist easy passage through thick low vegetation. Most of the 42 described species are associated with dense low groundcover such as spinifex, grass tussocks and complex heaths. As a secondary evolutionary move, some have subsequently adopted an underground lifestyle, reducing the size of the hind limb flaps and covering the ear-openings. Flap-footed lizards range from inconspicuous worm-like burrowing lizards that feed on ant larvae to surface-active predators nearly 1 metre long that prey on other lizards.

This is the only lizard family restricted to the Australian region, occurring in all areas except some cool temperate, alpine and wetland habitats. One widespread species reaches New Guinea and another is confined to that island. The south-western corner and lower west coast are the hotspots, supporting a large number of species including many endemics.

Burton's Snake-lizard: more like a snake than a lizard!

▲ A Burton's Snake-lizard (*Lialis burtonis*) consuming a gecko. This lizard specialist has a hinged skull to ensure a firm grip. Badu Island, Qld.

Australia's most widespread and variable lizard is also highly distinctive and extremely specialised. **Burton's Snake-lizard** (*Lialis burtonis*) is common from the south-west coast, across the continent to southern New Guinea, excluding only parts of southern Australia. It lives in dry or well-drained habitats, particularly areas with a thick grass or heath groundcover. The lizard is active by day or night, depending on temperature.

This unusual lizard is easily recognised by its long, pointed snout with a fine wedge-shaped tip. It is one of the largest flap-footed lizards, with a head and body length of nearly 30 centimetres and a total length approaching 1 metre. Burton's Snake-lizards come in an enormous range of colours, from brick red to bright yellow, brown, putty grey or almost white. Wherever they live, a number of colour forms occur together, but there are trends in the distribution of some patterns. Lizards with simple narrow stripes are common in spinifex deserts, those with broad black and white facial stripes are often seen in northern and eastern Australia, and plain grey specimens can turn up anywhere.

The Burton's Snake-lizard's resemblance to small snakes goes beyond mere outward appearance. They share with many small Australian snakes a specialised diet of small lizards. Burton's Snake-lizards are ambush predators, lying concealed in thick low vegetation ready to snatch passing prey with a sidewise swipe of the slender snout. The fine peg-like teeth are curved backwards to ensure a secure grip but the trump card is the uniquely hinged skull. It is flexible at about level with the eyes, allowing the tips of its upper and lower jaws to close and meet, completely encircling its prey. When the victim has been suffocated, the lizard manipulates its food into position and swallows it head-first. It looks like a snake and feeds like a snake but is a common and widespread lizard.

▲ A Burton's Snake-lizard (*Lialis burtonis*). They exhibit an extraordinary range of colour forms. Mt Tyson, Qld.

▼ Burton's Snake-lizard. Chesterton Range, Qld.

Loitering in the litter: worm-lizards and the Bronzeback

▲ The Flinders Worm-lizard (*Aprasia pseudopulchella*) favours heavy, stony soils in an area centred on the Flinders Ranges of South Australia. It is listed as Vulnerable.

If the flap-footed lizards developed long bodies and reduced limbs to exploit dense low vegetation, worm-lizards and the Bronzeback have gone one step further. Their ear-openings are completely covered or greatly reduced and the hind limb flaps are barely visible. These modifications suit their lifestyle in soil or under mats of thick leaf litter.

Worm-lizards

Worm-lizards (*Aprasia*) are sometimes seen active on the surface during the day, but most of their time is spent hidden in loose soil beneath rocks, logs and rotting stumps, in insect-holes and under leaf litter. The 12 named species are largely confined to southern Australia, from the Australian Capital Territory and central Victoria to the south-west, and along the mid-west coast.

Worm-lizards are aptly named, for these small lizards with a head and body length of 14 centimetres or less could easily be mistaken for wriggling worms. They have cylindrical bodies and their blunt tails are slightly shorter than their body length. All other flap-footed lizards have tails much longer than their bodies. Their snouts range from rounded to angular and beak-like, which probably reflects different burrowing lifestyles and substrates. Some species have sharply contrasting pink or black tails that may serve to divert a predator's attention to the 'disposable' end of the lizard. Their eyes are surprisingly large for lizards that tunnel through loose soil.

▲ Flinders Worm-lizard (*Aprasia pseudopulchella*). Burra area, SA.

Worm-lizards are often found in association with ant nests, where they probably dine on a smorgasbord of eggs, pupae and larvae. It is likely that ants make up the bulk or perhaps the entire diets of many species. Like all flap-footed lizards, they squeak when handled, and readily discard their tails.

The tail of the **Red-tailed Worm-lizard** (*Aprasia inaurita*) may act as a decoy, attracting a predator's attention away from the vulnerable body. It occupies woodlands and shrublands on sandy soils from north-western Victoria, across the Great

Australian Bight to the south-east coast of Western Australia.

Bronzeback

The **Bronzeback** (*Ophidiocephalus taeniatus*) was 'lost' for more than 80 years. It was named in 1897 from a specimen collected at Charlotte Waters in Central Australia and despite numerous searches by dedicated herpetologists no more were seen until the discovery of a population in 1978. In a story similar to that of the Pygmy Blue-tongue (see page 19), it took perseverance and a good dose of luck to unravel the mystery.

The Bronzeback has a shovel-shaped snout similar to some burrowing skinks, particularly sliders of the genus *Lerista*. This distinctive lizard, the sole member of its genus, lives under mats of leaf litter associated with the sparse vegetation lining watercourses in the bleak stony deserts of northern South Australia.

Now, armed with knowledge of their specialised requirements, Bronzebacks are better understood, and are being located over an increasingly large area. They are listed as Vulnerable.

▲ Red-tailed Worm-lizard (*Aprasia inaurita*). Port Germein, SA.

▶ Bronzeback (*Ophidiocephalus taeniatus*). Abminga, SA. D. Knowles

Whip-thin: delmas and the Keeled Legless Lizard

▲ Helmeted Delma (*Delma mitella*). Tolga, Qld.

Easy and speedy passage through a thick matrix of low vegetation demands extreme modifications. The 21 species of *Delma* are very smooth and slender with long tails, sometimes more than four times the body length. They have relatively large hind limb flaps and obvious ear-openings. Delmas thrive in spinifex deserts, dense heaths and areas with a groundcover including tussock grasses. Delmas are found throughout mainland Australia except some cool, moist southern areas.

Delmas are secretive lizards that rarely stray into the open, normally remaining within the protective cover of low dry vegetation. Within this shelter they slip swiftly through the tangled stems and foliage. If startled on open ground they move in a series of rapid leaps. Some delmas, particularly in southern Western Australia, commonly bask atop shrubby thickets. When disturbed they vanish in a blur so rapid one is left wondering if there really was a lizard there in the first place.

Some delmas have distinctive broad dark bands across the head and neck. Superficially this resembles the head pattern of juvenile brown snakes, though it is yet to be established that these harmless lizards are actually mimicking venomous snakes. It may be that both groups share a strategy of temperature management where a dark head absorbs heat rapidly without exposing the whole body to danger. Delmas are generalist invertebrate feeders, taking a variety of small insects such as cockroaches and grasshoppers.

▲ Striped Delma (*Delma impar*). Melbourne, Vic.

▼ Collared Delma (*Delma torquata*). Mt Crosby, Qld.

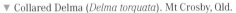

The **Helmeted Delma** (*Delma mitella*) is the largest delma, with a head and body length of 20 centimetres. This is a distinctive lizard, with a yellow belly, copper head bands and a thin dark stripe beneath each side of its tail. It has an unusual distribution, inhabiting the moist eucalypt forests that border tropical rainforests of north Queensland. Other delmas favour drier habitats.

The **Collared Delma** (*Delma torquata*) is the smallest species, reaching a head and body length of little more than 6 centimetres. It is restricted to south-east Queensland, favouring well-drained rocky slopes with tussock grasses and leaf litter. The most significant populations are threatened by exotic weeds and urban and agricultural development, so the species is listed as Vulnerable. Like many flap-footed lizards it is a sedentary species. Individuals are often located in association with one or more shed skins, indicating long-term residence in a shelter site.

The **Striped Delma** (*Delma impar*) is the most south-easterly occurring flap-footed lizard. It lives in temperate grasslands from the western edge of Melbourne to the Australian Capital Territory and adjacent New South Wales. Only a small amount of this unique habitat remains intact and it continues to decline in the face of rapid urban growth. The species is listed as Endangered.

121

▲ Javelin Legless Lizard (*Delma concinna*). Green Head, WA.

▼ Keeled Legless Lizard (*Pletholax gracilis*). Muchae, WA. B. Maryan

The **Javelin Legless Lizard** (*Delma concinna*) lives in complex heaths along the west coast. It is extremely swift and agile, usually encountered (very briefly!) basking in thick vegetation up to 1 metre or more above the ground.

The **Keeled Legless Lizard** (*Pletholax gracilis*) has no close relatives and is the sole member of its genus. Little is known about this secretive inhabitant of the complex heaths and banksia woodlands along the west coast. Two subspecies are recognised, a mid-western and a south-western form. It is shoelace-thin with an extremely long tail and two distinct keels on each scale around the body. The head and body length is about 8.5 centimetres. Though common in bushland around Perth, it is rarely seen active as it is very swift and the rough slender body is nearly invisible against a backdrop of twigs and dry foliage.

Scaly-foots: the snake-mimics

▲ Common Scaly-foot (*Pygopus lepidopodus*), striped form. Fitzgerald River NP, WA.

▼ Common Scaly-foot (*Pygopus lepidopodus*), plain form. Mt Nebo, Qld.

Scaly-foots are so named because their hind limb flaps are much larger and more obvious than those of other flap-footed lizards. Internally, their rudimentary limbs retain a near-complete assemblage of bones, and if handled the lizards will readily move them. In some respects, scaly-foots are the most conventionally lizard-like members of an otherwise highly divergent family.

These large, robust flap-footed lizards have mastered the art of copying snakes. When harassed they rear their heads and forebodies and rapidly flicker their tongues. Some species have the added bonus of dark heads resembling those of young brown snakes. A skilled naturalist would easily spot the obvious ear-openings and flat fleshy tongues – both typical lizard features – but the display is impressive enough to fool some predators.

Between them, the common and hooded scaly-foots (*Pygopus*) cover most of mainland Australia. They have non-glossy, matt-textured scales, and a row of eight or more pores in front of the vent. The three hooded species occupy the arid zones and dry tropics while the Common Scaly-foot occurs in southern and eastern areas. They prey on a variety of invertebrates, particularly ground-dwelling spiders.

The **Common Scaly-foot** (*Pygopus lepidopodus*) is the largest member of the family, reaching a head and body length of 27 centimetres and, in eastern Australia, a total length of nearly 1 metre. The scales on

123

▲ Eastern Hooded Scaly-foot (*Pygopus schraderi*). Moranbah, Qld.

its back and sides are strongly keeled. If handled, it rotates rapidly while uttering a loud squeak and will readily discard its tail. Large adults with complete original tails are uncommon. Normally at least the tip is regenerated. Common Scaly-foots are active by day in mild to warm weather, and at night in hot weather. They occur in heaths and grassy woodlands. Two distinct colour forms occur – a boldly striped form is mainly associated with southern heaths while plain forms occur throughout the range.

▼ Brigalow Scaly-foot (*Paradelma orientalis*). Chesterton Range, Qld.

The **Eastern Hooded Scaly-foot** (*Pygopus schraderi*) occurs in the dry eastern interior, mainly on heavy red soils. It has weakly keeled scales and a dark head patch that superficially resembles young brown snakes and some whipsnakes.

Like its relatives the Western and Northern Hooded Scaly-foots, this species is almost exclusively nocturnal.

The **Brigalow Scaly-foot** (*Paradelma orientalis*) is currently placed in its own genus. Its glossy scales have a smoky sheen and it has a row of four pores in front of the vent. This secretive nocturnal species occupies dry woodlands and outcrops in the eastern interior of Queensland. It feeds on a variety of invertebrates, and also climbs the trunks of rough-barked acacia trees to lick exuding sap. It is considered threatened by habitat fragmentation and is listed as Vulnerable. The individual here is rearing and flickering its tongue.

Dragons: lizards with attitude

▲ The Tawny Dragon (*Ctenophorus decresii*) inhabits rocky outcrops in southern South Australia and western New South Wales. Some males exhibit stunning breeding colours. Telowie Gorge, SA.

Thanks to their upright postures and armoury of spikes, crests, erectable frills and beards, dragons seem to strike a primeval chord, matching our notions of mythical monsters, and perhaps what dinosaurs may have looked like. They have dramatic appeal.

Dragons have rough scales and generally long limbs and tails. Most are very swift, and some can accelerate on their hind legs to reach mind-boggling speeds. A few are slow-moving but superbly camouflaged.

These alert lizards are strongly visually cued, so much of their behaviour, including social interactions and feeding strategies, is obviously influenced by keen eyesight. They frequently select elevated perches from which to survey each other and the surrounding terrain; males tend to develop clearly recognisable and often intense breeding and social colours; and dragons have a complex array of display signals including head bobs and dips, arm-waving and tail-lashing. Dragons are 'sit and wait' predators scanning from their chosen vantage points for the tell-tale movement of passing prey.

Dragons are related to the famous chameleons of Africa and Madagascar, and are the only Australian lizards that seize food with their tongues. But unlike the chameleons' high-speed projectiles, dragons

Boyd's Forest Dragons (*Hypsilurus boydii*) are 'sit and wait' predators that spy insects from their perches on upright trunks in north Queensland's Wet Tropics. Mossman Gorge, Qld.

dab their short thick tongues on a food item to pick it up. Small dragons feed exclusively on invertebrates – mainly insects – while larger ones are omnivorous. They browse on significant amounts of grass shoots, flowers and other vegetation. For some groups such as the sand and earless dragons, ants form a dominant portion of their diets, while Thorny Devils eat little or nothing else. These dragons must have devised means of tolerating unpalatable prey because other broadly insectivorous groups of lizards actively avoid ants, presumably due to the formic acids they contain.

All dragons are egg-layers, depositing their clutches in burrows. Egg deposition sites are often in surprisingly open areas, including windrows along the edges of tracks, on exposed dune slopes and even in a freshly delivered load of sand.

While several species occur in rainforests, along eastern watercourses, and just one, the Mountain Heath Dragon, reaches Tasmania, dragons are most successful in hot dry climates. The centre of diversity lies in Australian deserts and the seasonally dry tropical north. Many species are heat-tolerant, able to function efficiently with body temperatures well over 40° Celsius. In human terms this would be a dangerously high fever.

Dragons also employ colour and posture to adjust their temperatures. They can select dark tones and align their flattened bodies perpendicular to the sun to absorb heat, while pale hues and bodies angled into the sun reduce heat gain. They can also raise themselves high off hot substrates and rest on their claws and heels. As the mercury climbs and animals retreat to their shelters, it is often dragons that are the last lizards standing. They have truly mastered thermal management.

▲ A Southern Angle-headed Dragon (*Hypsilurus spinipes*) keeps a wary eye for predators. Mt Glorious, Qld.

▼ Looks can be deceiving. This fierce looking Thorny Devil (*Moloch horridus*) is a harmless Central Australian lizard. Simpson Desert, NT.

Ultimate bluff: Frilled Lizard and bearded dragons

▲ Frilled Lizard (*Chlamydosaurus kingii*). Kununurra district, WA.

One of the greatest tricks to foil predators is to look as fierce as possible. Suddenly expanding to nearly double size creates such a striking effect that Frilled Lizards and bearded dragons are among Australia's most famous and frequently illustrated reptiles.

Frilled Lizard

The **Frilled Lizard** (*Chlamydosaurus kingii*) is probably the most recognisable lizard in Australia. It appeared on the two cent coin, graces virtually every picture book on Australian wildlife and has achieved the status, alongside the kangaroo and koala, of an international fauna ambassador.

Frilled Lizards are common in tropical woodlands across northern Australia, but are difficult to find during the dry season when they spend most of their time in the trees. As the wet season arrives Frilled Lizards descend to the ground or position themselves at heights of 1–2 metres on rough-barked trunks. At these times they are a common sight along Top End road edges.

These lizards are probably Australia's largest dragons, with a head and body length of over 25 centimetres and a total length of nearly 75 centimetres. They are equipped with one of the most unusual appendages in the reptile world. When at rest, the thin scaly cape lying folded over their shoulders is inconspicuous, and if approached the lizard will crouch low to the ground or slide discreetly to the far side of a tree. But when threatened or provoked the pinkish yellow mouth is gaped and the frill dramatically erected to nearly encircle the head, standing out at right angles as a ribbed disc nearly 30 centimetres across. It is supported by several long, slender hyoid bones that can be raised and lowered as the mouth is opened. Some populations have vibrant splashes of red, orange or yellow on the frill. The lizard hisses and leaps toward its attacker until it spots the opportunity for a hasty retreat, sprinting on its hind limbs to the nearest tree.

While still abundant in northern parts of its range, the most south-easterly population near Brisbane has dramatically declined as a result of feral predators and habitat loss.

Bearded dragons

Bearded dragons (*Pogona*) have mastered the art of 'extreme bluff' by outwardly similar, yet subtly different means. Rather than a thin fold of skin, it is their throats that expand to form a beard, supported by shorter versions of the same hyoid bones. Bearded dragons are liberally adorned with spines, including a series along the edge of the beard, rows along their flanks, and a scattering over their backs. Their gaping mouths range from pink to bright yellow and when displaying, their bodies become pancake-flat and are tilted towards the aggressor to enhance the effect of increased overall size.

Between them, the six species cover most of Australia. Bearded dragons have

fared better than the Frilled Lizard in the face of development, with thriving populations in most capital cities and farm lands.

Bearded dragons seek elevated perching sites such as stumps and fence posts. From these vantage points they can spy passing prey and keep an eye on each other. In keeping with many large dragons, bearded dragons include significant amounts of vegetation in their diets. In addition to eating insects and other invertebrates, they are often seen in parklands, browsing on clover, dandelion flowers and other plants.

▼ Frilled Lizards (*Chlamydosaurus kingii*) spend much of the day clinging to tree trunks. Kununurra district, WA.

▲ Eastern Bearded Dragon (*Pogona barbata*). Dutton Park, Qld.

South Brisbane Cemetery is home to a thriving population of **Eastern Bearded Dragons** (*Pogona barbata*) with headstones offering perfect vantage points.

The displaying male (opposite) has been dining on vegetation, with some leaves still visible in its mouth. There is also evidence of tooth damage, a result of combat with another male. During the breeding season combating males grip each others' jaws, often resulting in damaged teeth and sometimes broken bones. This is the largest species of bearded dragon, with a head and body length of 25 centimetres.

On the other side of the continent and the opposite end of the size scale is the smallest of all, a subspecies of **Dwarf Bearded Dragon** (*Pogona minor minima*). This slender dragon with a poorly developed beard reaches a head and body length of only 11 centimetres. It is restricted to the Houtman Abrolhos islands off the west coast. Other slightly larger subspecies of *Pogona minor* extend from Western Australia to Central Australia.

Inland Bearded Dragons (*Pogona vitticeps*) resemble their large eastern relatives, but have rounder beards, broader heads, and the mouth is pink rather than yellow. This is the common species over vast tracts of Australia's eastern interior.

▲ Eastern Bearded Dragon (*Pogona barbata*), displaying. Dutton Park, Qld.

◄ Dwarf Bearded Dragon (*Pogona minor minima*). West Wallabi Island, WA. B. Maryan.

▼ Inland Bearded Dragon (*Pogona vitticeps*). Blackall, Qld.

City slickers: Water Dragons

▲ Mature male Eastern Water Dragon (*Physignathus lesueurii lesueurii*). Brisbane, Qld.

Huge populations of street-wise, habituated **Water Dragons** (*Physignathus lesueurii*) thrive in eastern Australian cities from Canberra to north Queensland. Who cannot be impressed by the sight of these large crested lizards, nearly 75 centimetres long, with powerful limbs and long, laterally compressed tails? They lie comfortably draped beside ornamental ponds in city gardens, and bask on rocks and logs near streams, rivers and duck ponds. Riverside restaurants are often frequented by resident water dragons that lurk under tables keeping a sharp eye out for fallen scraps and the odd handout.

There is just one species of water dragon in Australia, occurring as two distinct subspecies. The males of the southern race,

called **Gippsland Water Dragons** (*Physignathus lesueurii howittii*), have dark chests with blue and yellow blotches on the throat. They range from eastern Victoria to the Australian Capital Territory and across to about Kangaroo Valley, New South Wales. From there northward, the **Eastern Water Dragon** (*Physignathus lesueurii lesueurii*) takes over. These have a dark band behind the eye and a dull red chest.

Water dragons are powerful swimmers that seldom stray too far from water. In the cities, regular contact with humans has endowed them with a degree of nonchalance and they can easily be approached and sometimes handfed. (On a cautionary note, they can also slice thumbnails!) In a bushland setting the lizards are very wary and will not hesitate to leap into the water if

▲ Sub-adult Eastern Water Dragon (*Physignathus lesueurii lesueurii*). Brisbane, Qld.

▼ Mature male Gippsland Water Dragon (*Physignathus lesueurii howittii*). Wagga Wagga area, NSW. T. Annable

approached, sometimes dropping many metres from overhanging branches. When swimming the limbs are held to the sides and they are propelled forward by lateral undulations of the body and tail. They can slow their heartbeats under water and may remain submerged for up to two hours.

During spring, males engage in territorial combat. They begin by flattening their bodies, erecting crests and lowering dewlaps, before lying snout to snout and grasping each other by the jaws. Each male appears to dominate a harem of several females. Eggs are deposited in burrows dug in the creek bank and by mid-summer the slender young with their large bulbous heads can be seen basking on waterside vegetation. Adult water dragons are omnivorous but the young feed mainly on insects.

Eastern Water Dragons extend well into the Brisbane CBD. Along the banks of the meandering Brisbane River, they line the pedestrian and bike paths, paying little heed to the commuters hurrying past.

In parts of Victoria, Gippsland Water Dragons are known as salamanders, though they bear no resemblance to those tailed amphibians. The lizards will not hesitate to leap into cold flowing waters to escape capture.

Tommy round-heads and their kin

▲ Tommy Round-head (*Diporiphora australis*). Karawatha Forest, Qld.

The 14 species of tommy round-heads and their relatives (*Diporiphora*) are generally slender dragons with long limbs and tails. Some are so thin they appear emaciated bordering on starved. They range in size from a head and body length of 5 to 9 centimetres, and generally lack crests or other significant spines. Most species are patterned along a similar theme, with a pair of narrow pale stripes overlaying a series of short, broad dark bands. This is an ideal disruptive pattern to break up their outlines among twigs and low foliage. Some males seasonally acquire a prominent circular black blotch on each shoulder.

The group is distributed across northern Australia and down the east coast to northern New South Wales, with outlying species in the sandy deserts. These dragons bask among low vegetation such as spinifex, shrubs and fallen branches. A few perch prominently on elevated sites such as termite mounds. Most are not particularly swift compared to other dragons, scuttling on all four limbs to cover when disturbed.

The various species are often difficult to identify and it is likely that many are unnamed, particularly in parts of Queensland where very little study has been undertaken. It is often necessary to check for discrete skin folds beneath the throat, behind the ear and across the shoulder to try to distinguish them.

The smallest member of this group is the enigmatic *Diporiphora convergens*. Just one individual is known, an immature specimen with a head and body length of less than 3.5 centimetres found at Admiralty Gulf in

◄ Superb Dragon (*Diporiphora superba*). Manning Creek, WA.

▼ Tommy Round-head (*Diporiphora australis*). Townsville area, Qld.

the northern Kimberley region of Western Australia in 1972. It has never been photographed.

The **Superb Dragon** (*Diporiphora superba*) of far northern Western Australia is built more like a stick insect than a lizard. Its body and limbs are so thin and the tail so long the animal is ungainly on the ground. It is more comfortable among the foliage of low shrubs, where its shape and green colouration provide effective camouflage. This may be Australia's most consistently arboreal dragon.

The **Tommy Round-head** (*Diporiphora australis*) is common along the east coast, including bushland around Brisbane. It lives in heaths and eucalypt forests, where it is often seen perching among thick low vegetation beside tracks.

The **Arnhem Two-lined Dragon** (*Diporiphora arnhemica*) is a tropical species inhabiting low open woodlands with a groundcover of spinifex. It often perches conspicuously on rocks or termite mounds. This alert dragon seems much swifter than many of its relatives.

▲ Arnhem Two-lined Dragon (*Diporiphora arnhemica*). Halls Creek, WA.

Comb-bearing dragons: rock-hoppers, sleek speedsters and round-headed burrowers

▲ Ring-tailed Dragon (*Ctenophorus caudicinctus*). Cathedral Gorge, WA. B. Maryan

The 23 members of the genus *Ctenophorus* are a diverse group ranging from slender, extremely swift species that accelerate across open spaces between spinifex clumps to stout, deep-headed lizards perching prominently on elevated sites near their burrows.

Species tend to form clusters based on shared physical features and lifestyles, but between them they occur in all arid areas where, sitting on roadside vantage points, they are sometimes the most conspicuous reptiles. They also extend to seasonally dry, well-drained rocky habitats in some southern and western areas.

Ctenophorus probably derive their name from a serrated fringe along their eyelids. They also share a distinctive row of enlarged scales curving under each eye. Mature males usually develop breeding colours, and acquire varying amounts of black pigment on the chest and throat. Males also have an obvious row of pores along the underside of the thigh and in front of the vent. All species are keen-eyed, alert lizards, quick to sprint for safety or dive into a burrow or rock crevice if approached.

Rock-hoppers

The eight rock-inhabiting species of *Ctenophorus* dwell on outcrops across dry areas of the continent between western Queensland and the south-west of Western Australia. Most are obligate rock-specialists that rarely venture onto surrounding terrain. This has resulted in considerable genetic isolation, and many species feature distinctive colour variants that occupy different outcrops with little exchange between them. Differences mostly relate to males; in many cases there is little colour difference on females from separate outcrops or even between females of some closely related species.

Most rock *Ctenophorus* feature brightly-hued males that use protruding rocks and boulders as display sites. From these vantage points they exhibit coloured flanks and bob their heads to flash their throats, signalling sexual status and condition using colours in the UV spectrum.

Ring-tailed Dragons (*Ctenophorus caudicinctus*) are the most widespread rock dragons, extending from western Queensland, through Central Australia to cover most of the Western Australian interior. Six distinct subspecies are named across this broad range. Males of some races develop sharply contrasting black rings on their tails.

Red-barred Dragons (*Ctenophorus vadnappa*) live in the northern Flinders Ranges of South Australia. Females are mottled shades of brown but mature males have pastel blue backs and black and red slashed flanks. At the sight of rival males they rear high on all four limbs, extend their striped yellow to orange throats, do push-ups and coil their tails vertically over their backs.

'Islands' of exposed granite in the south-west of Western Australia are home to the **Ornate Dragon** (*Ctenophorus ornatus*). It lives in areas where sheets of flat rock are strewn with thin exfoliated slabs. The lizard has an extremely depressed head and body, allowing it to sprint into narrow crevices when pursued. Distinct colour forms occur on isolated outcrops.

The **Yinnietharra Rock Dragon** (*Ctenophorus yinnietharra*) is restricted to a single pastoral station in the western interior of Western Australia. Throughout

▲ Red-barred Dragon (*Ctenophorus vadnappa*). Blinman Creek, Flinders Ranges, SA.

▼ Ornate Dragon (*Ctenophorus ornatus*). Gibb Rock, WA. B. Maryan

▲ Yinnietharra Rock Dragon (*Ctenophorus yinnietharra*). Yinnietharra Station, WA.

the midday heat the dragons perch on protruding rocks, supported on the heels of their feet, with bodies held high off the hot substrate.

Sand dragons

The five species of sand dragons range from the mallee regions of western Victoria and New South Wales to the southern Kimberley region of Western Australia. Sometimes referred to as military dragons, these slender, slightly flattened, long-limbed lizards are among the speediest animals in the desert. Their realm is the open spaces between low arid vegetation such as spinifex and shrubs.

Sand dragons seem to spend their lives in near-perpetual activity, running on all four limbs from clump to clump, snapping up ants and keeping an eye out for danger. If approached they generally dash a few metres and pause, knowing they have the capacity to sprint beyond reach of any attempt to seize them. Alas, life in the fast lane comes at a price. At least one species is known to be annual, with an almost complete population turnover each year. Unlike other *Ctenophorus*, these dragons rarely perch on raised objects, preferring to view their world and each other from ground level.

▲ A female Central Military Dragon (*Ctenophorus isolepis*). Hay River, NT.

▲ A male Central Military Dragon (*Ctenophorus isolepis*). Ewaninga, NT.

◄ Mallee Dragon (*Ctenophorus fordi*). Menzies, WA.

The **Mallee Dragon** (*Ctenophorus fordi*) lives in semi-arid mallee and spinifex communities across southern Australia. Thanks to their sleek bodies and long slender limbs, they can out-sprint most predators but few individials live longer than one year.

The **Central Military Dragon** (*Ctenophorus isolepis*) has a huge distribution. There are three subspecies which, between them, span most of the arid zone. Visitors to Central Australia's peak tourist attraction, Uluru, often see this racy lizard darting between spinifex clumps.

The **Sandhill Dragon** (*Ctenophorus femoralis*) inhabits bare desert dune slopes and crests on the mid-west coast in the vicinity of North West Cape, Western Australia. The surrounding sandy flats are the domain of the Central Military Dragon.

▲ Sandhill Dragon (*Ctenophorus femoralis*). Bullara Station, WA.

Bicycle dragons

When sprinting, many dragons raise their bodies and run on their hind legs with forelimbs hanging loosely and the long tail streaming back as a counterweight. The pedalling motion has earned them the name bicycle lizards. The title is often applied more specifically to two very speedy, long-limbed species, the Crested and Lozenge-marked Bicycle Dragons, from semi-arid southern woodlands and shrublands. A third closely related species, McKenzies' Bicycle Dragon, is poorly known but probably also employs this bipedal gait.

The **Crested Bicycle Dragon** (*Ctenophorus cristatus*) extends from the Eyre Peninsula to southern Western Australia. It has a crest of large spines on its neck and a smaller crest along each shoulder, intense black and cream to orange marbling on the forebody, and striking black rings around its tail. It is terrestrial, but selects perching sites on low fallen timber.

The **Lozenge-marked Bicycle Dragon** (*Ctenophorus scutulatus*) occurs in the Murchison and Goldfields regions of Western Australia, often in acacia woodlands on heavy soils.

Round-headed burrowers

Though generally not as swift as other *Ctenophorus* the burrowing species are no less wary. The eight species mostly shelter in short sloping burrows, excavated at the bases of bushes or stumps. From nearby vantage points they survey their territories and scan for food, but will dash into their burrows at the first sign of danger.

These are generally stout dragons with relatively short limbs and tails and deep heads with short snouts. Included within this group are some widespread generalist species that between them occupy virtually the entire Australian arid zone, and some specialised inhabitants of salt lakes and stony plains. Most, probably all, species have a high heat tolerance.

The **Central Netted Dragon** (*Ctenophorus nuchalis*) is a distinctive 'golfball-headed' dragon commonly seen basking on windrows, termite mounds and discarded tyres along the edges of outback roads. They retreat to their burrows during temperature extremes and backfill them with a soil plug during winter inactivity. It is the most widespread member of the genus.

◀ Crested Bicycle Dragon (*Ctenophorus cristatus*). Lake Hurlestone, WA. D. Knowles

▲ Lozenge-marked Bicycle Dragon (*Ctenophorus scutulatus*). Wooleen Station, WA.

▲ Central Netted Dragon (*Ctenophorus nuchalis*). Hay River, NT.

▲ Lake Eyre Dragon (*Ctenophorus maculosus*). Lake Eyre North, SA.

The surface of Lake Eyre in northern South Australia is so harsh that the **Lake Eyre Dragon** (*Ctenophorus maculosus*) has the place all to itself. No other vertebrate lives on the buckled salt crust, devoid of all vegetation and fresh surface water. The only available food is the harvester ant, plus insects windblown from adjacent dunes. The salt crust overlies a layer of fine, moist, silty sand. This is the lizards' retreat when temperatures soar and duststorms howl. Males stake out territories surrounding any raised vantage point – a large chunk of salt, a piece of driftwood or even an animal skull. They relentlessly pursue females who must sometimes resort to flipping onto their backs to avoid unwanted amorous attention. During rare floods, the entire lizard population must evacuate the lake and take up residence on sand dunes beyond the shore. There they dominate and bully the resident Painted Dragons.

Painted Dragons (*Ctenophorus pictus*) occupy arid to semi-arid southern sandy areas including heaths, spinifex and cane grasses. Favourite burrow sites are the consolidated mounds of soil at the bases of some desert bushes. Though populations differ, breeding males acquire varying amounts of blue, yellow and orange pigments.

▼ Lake Eyre Dragon (*Ctenophorus maculosus*). Lake Eyre North, SA.

142

▲ Painted Dragon (*Ctenophorus pictus*) male. Moomba area, SA.

▼ Painted Dragon (*Ctenophorus pictus*) female. Currawinya NP, Qld.

In a world of shadows: rainforest dragons

▲ Juvenile Southern Angle-headed Dragons (*Hypsilurus spinipes*) have angular brows resembling dead leaves. Mt Glorious, Qld.

The rainforest dragons (*Hypsilurus*) are unusual as much for their choice of habitat as for their striking appearance. Most Australian dragons inhabit dry, open sunny areas but the two rainforest or angle-headed dragons live in muted light in widely separated subtropical and tropical rainforests.

Both species have impressive crests of large triangular scales on their napes, strongly laterally-compressed bodies, long thin limbs and acutely angular brows. Ours are endemic Australian species but they are closely allied to numerous New Guinea lizards. Their origins hark back to times when very extensive and complex tropical forests linked large areas of Australia with New Guinea.

It takes a keen eye to spot rainforest dragons perched on the upright trunks and stems of trees, saplings and vines. These slow-moving lizards generally slide discreetly from view if approached, rather than drawing attention to themselves with a noisy dash for cover. When resting motionless, their spines and angular, ragged outlines help camouflage the dragons against a background of dappled light and contrasting shapes. They spend long periods of time sitting quite still before suddenly leaping from their perch to seize a passing worm or other invertebrate. They also take small vertebrates but there is no indication of the partial herbivory that is common in other moderate to large dragons. Rainforest dragons have head and body lengths of 11–16 centimetres.

Boyd's Forest Dragon (*Hypsilurus boydii*) occurs in the Wet Tropics of north

◀ Boyd's Forest Dragon (*Hypsilurus boydii*). Mossman Gorge, Qld.

▼ Southern Angle-headed Dragon (*Hypsilurus spinipes*). Mt Glorious, Qld.

Queensland. It is easily recognised by the large plate-like scales on its jaw and the serrated line of sharp scales along the leading edge of the dewlap. This dragon rarely basks. It operates efficiently at very variable temperatures, between about 19–30° Celsius, equivalent to ambient air temperatures. For this slow-moving 'sit and wait' lizard, it appears that chasing errant puddles of sunlight through the rainforest takes too much effort.

The **Southern Angle-headed Dragon** (*Hypsilurus spinipes*) lives in the subtropical forests of north-eastern New South Wales and south-eastern Queensland. It has a smaller crest, and lacks the large scales on the jaw. Normally very difficult to find, these dragons can sometimes be encountered on open ground during early summer, when females excavate nest holes to lay their eggs near walking and vehicle tracks.

Devil of a lizard: Thorny Devil

▲ Thorny Devil (*Moloch horridus*). Hay River, NT.

The **Thorny Devil** (*Moloch horridus*) cannot be confused with any other animal. The dumpy body, short limbs and tail are covered with large thorny spines that render the animal unswallowable by most predators. It also has a horn-like pair of spines above each eye and a bulbous 'false head' complete with horns on its nape. The striking pattern of sharply contrasting black, white, yellow and ochre blotches and stripes are surprisingly effective at breaking the animal's outline against dappled shade under desert shrubs and spinifex. There is nothing remotely like it.

Thorny Devils live on sandy flats and dunes across vast tracts of the central and western deserts, extending into semi-arid heathlands on the west coast and south-western interior. They are normally encountered on roads, where the upcurved tail and slow, jerky gait like a mechanical toy are a distinctive sight. When disturbed they often freeze in mid-stride, even with a leg poised in mid-air.

The lizards have a highly specialised diet and a unique way of feeding. They consume only small black ants, positioning themselves above an ant trail so each ant passes under their nose. With repeated dips of the head and dabs with the short thick tongue, they dine on a meal of 1000 or so at a time. Each lizard visits a regular defecation site, where examination of scats reveals few sand grains among those crunched ant remains. All those direct hits on scurrying ants and almost no near misses!

▲ With grim determination, an amorous male Thorny Devil clings to the back of a larger female. Alice Springs, NT.

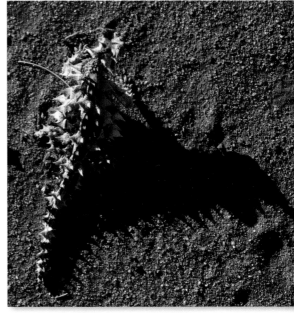

▲ Thorny Devil (*Moloch horridus*). Hay River, NT.

Thorny Devils also drink in an unusual way. Between the large spines the scales are fine and granular. Microscopic channels between these tiny scales lead water to the corners of the mouth. They can apparently drink if any part of the body is contacting water. Standing with one foot in a puddle a lizard can lick its lips to take in water.

When the Thorny Devil was first named in 1841 it was given the names *Moloch*, after a dreadful Canaanite god to which children were sacrificed, and *horridus*, meaning primarily 'rough' or 'bristly' but secondarily 'dreadful'. Clearly British zoologist Dr John Edward Gray was less than impressed with this extraordinary lizard.

The cryptic ones: Chameleon, Mulga and Pebble dragons

▲ Chameleon Dragon (*Chelosania brunnea*). Jabiru, NT. P. Horner

For some dragons, survival lies in concealment through camouflage or extremely secretive behaviour. Not surprisingly, such lizards are poorly known and infrequently seen.

Chameleon Dragon

The **Chameleon Dragon** (*Chelosania brunnea*) of northern Australia is rarely encountered crossing roads and is almost never seen at other times. It remains a poorly known reptile, despite its wide range from the Kimberley region of Western Australia to far north-western Queensland.

Like the true chameleons of Africa and Madagascar, it has a laterally compressed body, its eyes are largely enclosed within conical turrets of scaly skin and its blunt-tipped tail is slightly prehensile. If provoked the Chameleon Dragon distends a large dewlap.

The Chameleon Dragon lives in tropical eucalypt woodlands. It is arboreal, probably dwells largely in the canopy, and moves through branches slowly and deliberately, never drawing attention to itself. It is said to be clumsy, having been recorded falling from trees on several occasions. These lizards descend only rarely, with males recorded on the ground during May and females seen laying eggs in the mid-dry season between July and August. Regular sightings of individuals suggest they occupy permanent territories.

Chameleon Dragons are not listed as of conservation concern. Though poorly known and rarely seen, they may simply be 'difficult to find'. That said, it is likely that late dry-season fires and feral cats represent significant threats.

Mulga Dragon

In the arid southern interior of Western Australia, crisp dry twigs and fallen branches accumulate among the mulga bushes. Against this backdrop the cryptic little **Mulga Dragon** (*Caimanops amphiboluroides*) is virtually invisible.

This grey-streaked lizard, the sole member of its genus, has a rough texture formed by five crests of enlarged scales running down its back. It has short limbs, a blunt-tipped tail, and grows to a head and body length of about 9 centimetres.

The lizard does not betray itself with sudden movements. An ambush predator, it perches motionless on a grey log, darting forth if an insect strays near, and freezing or sliding from view if danger approaches. It takes a keen eye to spot one. During cool weather they have been uncovered beneath leaf litter.

Pebble Dragon

Western Australia's southern Kimberley region features semi-arid spinifex grasslands and open woodlands that stretch across into the adjacent western Northern Territory. Between the spinifex clumps, the ground is exposed and stony. Against this background, the **Pebble Dragon** (*Cryptagama aurita*) has evolved one of the most superb reptilian examples of camouflage through direct mimicry.

The small dumpy lizard mirrors the surrounding reddish brown pebbles. In fact, with a round head and a rotund body it copies two pebbles. Its blunt tail is shorter than the body, a rare feature in dragons.

Almost nothing is known of this small dragon. Less than half a dozen have been found and the animal pictured is the only individual ever photographed. The

▲ Mulga Dragon (*Caimanops amphiboluroides*). Paynes Find area, WA. B. Bush

superficially similar earless dragons (*Tympanocryptis*) also feature artful pebble-mimics (see page 154). If those are anything to go by, this lizard probably feeds mainly on ants and when disturbed would tuck its short limbs close to the body and nestle among the stones rather than flee.

▲ Pebble Dragon (*Cryptagama aurita*). This is the only one of its species known to have been photographed alive. Wave Hill, NT. P. Horner

In the shrubbery: heath dragons

▲ Mountain Heath Dragon (*Rankinia diemensis*). Anglesea, Vic.

The three species of heath dragons (*Rankinia*) are dumpy little lizards with relatively short limbs and tails, and head and body lengths of 4.5 to 8 centimetres. They have low spines scattered over their bodies, a row of larger spines along each side of the tail base, and patterns that include rows of angular to hourglass-shaped blotches down their backs.

Heath dragons live along sunny forest edges and heaths across southern Australia, and among semi-arid shrublands on the mid-west coast. They rarely perch on raised objects, though eastern Australia's Mountain Heath Dragon often views its surroundings from a low rock or fallen branch. They are not particularly swift compared to other dragons, and when approached they usually freeze in the hope their disruptive angular patterns will conceal them. If pursued they scuttle on all four limbs to the cover of thick low vegetation.

The **Mountain Heath Dragon** (*Rankinia diemensis*) is the world's most southerly dragon, and the only member of the family to reach Tasmania. Much of its mainland habitat, between southern Victoria and the New England Tablelands of New South Wales, lies in upland areas subject to winter snowfalls. It is common and secure over most of its wide range, but several Victorian populations, including the lizard pictured from coastal heath at Anglesea, are isolated and considered to be threatened.

▲ Western Heath Dragon (*Rankinia adelaidensis adelaidensis*). Jandakot, WA.

▼ Shark Bay Heath Dragon (*Rankinia parviceps butleri*). Tamala Station, WA.

There are two subspecies of the **Western Heath Dragon** (*Rankinia adelaidensis*). *R. adelaidensis adelaidensis* lives in the heaths and banksia woodlands on the coastal sand plains around Perth. It has dark stripes on its belly, while *R. adelaidensis chapmani* from south coastal Western Australia and South Australia has irregular dark blotches.

The two subspecies of heath dragon (*Rankinia parviceps*) live on the mid-west coast. They are among Australia's smallest dragons, with head and body lengths of only 4.5 centimetres. They differ from other heath dragons in having their ears covered by scaly skin, and less angular dorsal markings. The pale, weakly patterned Gnaraloo Heath Dragon (*R. parviceps parviceps*) lives between Carnarvon and North West Cape, while the more southerly **Shark Bay Heath Dragon** (*R. parviceps butleri*) lives between Shark Bay and Kalbarri. It is darker, with a bright yellow patch on its chin.

On the open plains: earless dragons

▲ The Four-pored Earless Dragon (*Tympanocryptis tetraporophora*) inhabits vast tracts of desert throughout across inland Australia. Parachilna, SA.

It comes as some surprise when driving through arid, treeless stony plains to see small dragons perched on roadside rocks, fully exposed to the sun in seemingly lethal temperatures.

The eight species of earless dragons (*Tympanocryptis*) live in open habitats across Australia, from temperate grasslands in Victoria, the Australian Capital Territory and New South Wales to some of Australia's harshest stony deserts.

Earless dragons are so named because the ear-drum or tympanum is completely covered by scaly skin and underlying muscle. They may actually be deaf. They are small, squat dragons with head and body lengths around 5 to 7 centimetres, relatively short limbs, and small spines scattered or arranged in rows over their bodies. All species are well camouflaged, with colours that closely match the red to brown substrates. Some are virtually patternless while others have bold disruptive patterns that feature narrow pale stripes overlying broad dark bands. Some species are extraordinary pebble mimics, with round heads and bodies to mirror the gibber stones they inhabit

Earless dragons may be extremely heat tolerant, but they also adopt postures to minimise thermal overload. Some species stand on their hind limbs, using their tails as tripods. By angling their bodies directly into the sun they can reduce the amount of exposed surface area while their white bellies reflect radiant heat from below. On open plains where the largest surface objects are no bigger than house bricks the upright posture also affords an improved view of their surroundings. When temperatures climb too high, they retreat down soil cracks or into burrows under rocks.

Earless dragons feed exclusively on small invertebrates, with ants forming a large part of their diets.

The **Grassland Earless Dragon** (*Tympanocryptis pinguicolla*) is listed as Endangered. Once common in temperate grasslands west of Melbourne, it has not been seen in Victoria for over 40 years and is probably regionally extinct. Populations persist in the Australian Capital Territory and adjacent New South Wales. The lizard pictured belongs to a population being studied along grassy road verges and in cotton and sorghum crops in Queensland's Darling Downs. Their initial identification as Grassland Earless Dragons is tentative, with genetic studies linking them to the Four-pored Earless Dragon. They may even represent an unnamed species. The secretive lizards normally scuttle from view or nestle against the soil, but the dragon pictured is employing a spirited display of bluff.

The **Blotch-tailed Earless Dragon** (*Tympanocryptis cephalus*) is a striking pebble mimic that lives in gibber deserts from the interior of Western Australia to western Queensland. It can sometimes be seen perching on roadside rocks but when disturbed the little lizard tucks in its legs, shuffles its body among the stones, and miraculously disappears.

▲ 'Grassland Earless Dragon' (*Tympanocryptis* species). Bongeen, Qld.

▼ Blotch-tailed Earless Dragon (*Tympanocryptis cephalus*). Twin Peaks Station, WA.

Sitting up like Jacky: the lashtails

▲ Jacky Lizard (*Amphibolurus muricatus*). Storm King Dam, Qld.

These swift, semi-arboreal dragons perch on trunks, branches, logs and rocks. They are keenly alert and quick to flee if approached, dashing on all fours or sprinting on their hind limbs.

The seven species of lashtails (*Amphibolurus*) have a low crest of spines on the neck and back, and patterns dominated by a pair of pale stripes or two rows of blotches. They have long limbs, and their tails range from moderate to extremely long and whip-like. Lashtails are medium-sized dragons, with head and body lengths of 8 to 13 centimetres.

Lashtails fall into two broad, ill-defined groups. The species occurring in southern and eastern Australia are largely associated with dry open forests and woodlands, while those in the centre and north are most common along watercourses. The latter group are sometimes referred to as *Lophognathus*. All species eat a variety of invertebrates plus small vertebrates such as skinks.

The **Jacky Lizard** (*Amphibolurus muricatus*) is common in eucalypt forests of south-eastern Australia. The grey colours and rough texture conceal it well as it perches immobile on rocks and stumps, but when males sight each other they draw attention to themselves with a series of tail-flicks, followed by arm-waves and push-ups. It is believed that the display varies to indicate a dominant or submissive status. When cornered and threatened the lizards gape their mouths to reveal a bright yellow interior.

155

◄ Nobbi Dragon (*Amphibolurus nobbi nobbi*). Girraween NP, Qld.

▼ Swamplands Lashtail (*Amphibolurus temporalis*). Saibai Island, Qld.

There are two subspecies of **Nobbi Dragon** (*Amphibolurus nobbi*). The northern form, *A. nobbi nobbi*, is common in dry forests and outcrops along eastern Australia, north of the Warrumbungle Mountains, New South Wales. *A. nobbi coggeri* occupies drier areas south and west of those ranges. Females of both races tend to have pale grey stripes broken or notched by black blotches but males, like the northern form pictured, develop lemon-yellow stripes and a mauve flush on the tail-base.

Gilbert's Dragon (*Amphibolurus gilberti*) is abundant across northern Australia where it is a familiar lizard in tropical gardens. It is often known colloquially as the Ta-ta Lizard, because arm-waving is a common feature of its display. The sharp black and white colours are typical of mature males. Females are more muted shades of grey.

The **Swamplands Lashtail** (*Amphibolurus temporalis*) occurs along the edges of watercourses across northern Australia, extending through Torres Strait to southern New Guinea.

▼ Gilbert's Dragon (*Amphibolurus gilberti*). Kununurra, WA.

Mammal-like intelligence: monitors

▲ Large adult monitors like the Perentie (*Varanus giganteus*) have outgrown all native predators and developed their own special kind of attitude. North West Cape, WA.

When a large monitor with sagging belly and scarred hide strides into a picnic ground it is greeted with fascination and much finger-pointing, and often a small degree of alarm. We wonder how we should react if it comes too close to our food and decides to help itself, and lurking in the backs of our minds are those tall tales of startled monitors running up a man's body. That combination of immense size, swaggering walk and apparent indifference present a primitive reptilian demeanour with which we are not always entirely comfortable.

With 27 named species, Australia is home to about half of the world's monitor lizards. The remainder range from South-east Asia to Africa and the Middle East. Australians generally call them goannas, a name borrowed and modified from the South American lizards, the iguanas.

All monitors are classified in the genus *Varanus*. They have loose granular skin, long snouts and necks, powerful limbs, strong claws and deeply forked, constantly flickering tongues. Their tails range from thick spiny appendages that block burrow entrances to slender, prehensile grasping tools. For some larger species, a muscular, laterally compressed tail is a formidable defensive weapon.

About nine Australian species reach 1 metre or more and include some of the world's largest lizards. The Perentie (*Varanus giganteus*) exceeds 2 metres and the Lace Monitor (*Varanus varius*) approaches 2 metres. By any reckoning these are enormous reptiles. A uniquely

Australian group, characterised by tails which are round rather than laterally compressed in cross-section, has evolved dwarfism. They reach their extreme in the world's smallest monitor, the 23 centimetre Short-tailed Monitor (*V. brevicauda*). These pygmy monitors are often classified in a separate subgenus, *Odatria*.

Monitor lizards are the largest native, land-based predators and scavengers in Australia. This places Australia in the unique position as the only continent where highly mobile, predatory reptiles still hold their place on the pointy end of the food pyramid. Elsewhere, mammals such as felines, hyenas and canines occupy that role.

As they hunt, monitors use their deeply forked tongues to collect airborne chemical data, transferring it to an organ in the roof of the mouth for analysis. They are the only Australian lizards to have tongues modified exclusively for a sensory role, playing no part in food manipulation. The organ is probably the most acutely sensitive tracking device among lizards. Because it is forked, the lizard has a direction-seeking capacity, with differences in particle loads on each lobe allowing it to follow mates and prey, and determine the source of carrion.

Some goannas capture animals by foraging, often covering several kilometres as they investigate burrows, detect scents and survey their surroundings with a keen eye. Others prefer ambush, darting out from concealed vantage points to seize passing prey. Their slender recurved teeth are designed for grasping rather than cutting or chewing. This means if the prey cannot be

▲ The Spotted Tree Monitor (*Varanus scalaris*) belongs to a complex of small arboreal species found across northern Australia. Burdekin River, Qld.

▲ The Spencer's Monitor (*Varanus spenceri*) is a sturdy lizard confined to treeless grassy plains on cracking clay soils. Central Qld.

ripped apart with teeth and claws it must be swallowed whole, consumed with jerking movements of the head and neck.

Some researchers have attributed monitors with mammal-like intelligence. It makes sense that high-order predators can predict prey behaviour and respond accordingly. And as they sometimes cover large areas through harsh terrain, it is important to be armed with a comprehensive mental map of the surrounds. Captive observations and experiments indicate that monitors recognise individual keepers, exhibit curiosity and can even count. Never underestimate a goanna!

▲ By rearing to gain a better view and 'tasting' the air for scent particles, a Gould's Monitor (*Varanus gouldii*) has a clear visual and chemical map of its surroundings. Glenmorgan, Qld.

159

Wedged in tight: ridge-tailed monitors

▲ Spiny-tailed Monitor (*Varanus acanthurus*). Kununurra, WA.

Thanks to their stout tails armoured with spiny scales, the ridge-tailed monitors can wedge themselves securely into burrows, rock cracks and cavities under termite mounds. Then, by blocking the entrance with their tails and inflating their bodies with air, there is little most predators can do to remove them. The four species of ridge-tailed monitors are restricted to northern Australia, mainly in seasonally dry areas featuring outcrops, stony ridges or hard soils with large termite mounds.

Ridge-tailed monitors are extremely secretive lizards. They are usually found hidden within their retreats rather than active on the surface, suggesting they are mainly 'sit and wait' predators, and that much more time is spent resting than foraging. They are terrestrial lizards and can sometimes be seen perched on low rocks or termite mounds. Some species occur in high population densities.

The **Spiny-tailed Monitor** (*Varanus acanthurus*) is the largest and most widespread species, with a total length of about 70 centimetres and a range encompassing western Queensland, the

Northern Territory and the northern half of Western Australia. It is variable in appearance, but typically features numerous dark-centred cream circles over its back, bands on its tail and longitudinal stripes on its neck. A northern subspecies, *V. acanthurus insulanicus*, restricted to islands off the north of the Northern Territory, is larger and darker with a banded pattern. Grasshoppers feature prominently among prey items recorded.

Storr's Monitor (*Varanus storri*) tends to favour low weathered outcrops, digging shallow U-shaped burrows under embedded rocks. Two subspecies occur, with *V. storri storri* in the north-eastern interior of Queensland and *V. storri ocreatus* between western Queensland and northern Western Australia. Lizards often occur in colonies of high population density but there is evidence to suggest these decline sharply wherever the introduced Cane Toad (*Bufo marinus*) is present.

▲ Storr's Monitor (*Varanus storri storri*). Charters Towers, Qld.

Secret lizard business: the pygmy burrowing monitors

▲ Desert Pygmy Monitor (*Varanus eremius*). Giralia, WA.

Some of the most secretive monitors are the pygmy burrowing species from the vast spinifex deserts. In the soft red sand, the prints of countless desert dwellers are laid bare, telling many tales to those who can read them. Among the dotted trails of centipedes and the meandering squiggles of slider skinks are the distinctive tracks of pygmy monitors, with their tail drag-marks clearly edged by the prints of small feet.

The **Desert Pygmy Monitor** (*Varanus eremius*) is extremely wary and seldom seen, yet its tracks indicate it forages widely over an extensive home range, investigating all fresh diggings as it hunts invertebrates and small lizards. This small monitor, with a total length of only 45 centimetres, excavates its own burrow and also makes use of those dug by other lizards. It appears to have an uncanny memory of where the burrows lie, as it will dash directly to the nearest if confronted. It is a strongly visually-cued hunter, using both ambush and swift pursuit to capture prey.

The **Short-tailed Monitor** (*Varanus brevicauda*) is the world's smallest monitor species, with a total length of just 23 centimetres. It is easily recognised by its short, thick tail, long body and short limbs.

Trapping programs on sand plains with long-unburned spinifex clumps reveal it to be common, yet this secretive little monitor is rarely encountered. Much of its prey consists of insects and many of these may be captured within the spinifex. It rarely ventures beyond the protective shelter of spinifex spines and the majority of the small lizards it takes are probably captured by ambush and brief forays.

▲ Short-tailed Monitor (*Varanus brevicauda*). Central Australia.

A life in the trees: small arboreal monitors

▲ Bush's Pygmy Monitor (*Varanus bushi*). Mt Robinson, WA. B. Bush

Hollow limbs and cavities behind loose bark make excellent retreats for small monitors. Most timbered areas over the northern two-thirds of Australia harbour at least one, and in some localities two or three, species of well-clawed arboreal goannas that scale the trunks to investigate nooks and crannies for insects, small lizards and birds' nests. For at least six species of small monitors, trees provide the primary shelter sites and hunting grounds.

The secretive **Bush's Pygmy Monitor** (*Varanus bushi*) is a recently described species, named in 2006. This lizard, with a total length of only 35 centimetres, occurs in the Pilbara region of Western Australia. In much of arid Western Australia it is replaced by the closely related Stripe-tailed Monitor (*V. caudolineatus*) and in Central Australia by Gillen's Pygmy Monitor (*V. gilleni*). Bush's Pygmy Monitor is the most poorly known of the trio but they are likely to have similar habits. The Stripe-tailed and Gillen's Pygmy Monitors are infrequently seen active, yet can be locally abundant and easy to find behind dead bark in stands of mulga and desert oak trees. They forage early and return to their retreats as temperatures rise. They are recorded to prey on both terrestrial and arboreal geckos, plus invertebrates such as large insects and

▲ The Black-headed or Freckled Monitor (*Varanus tristis*) is a wide-ranging species that frequently inhabits rocky outcrops and overhangs as well as trees. Windorah, Qld.

scorpions. Geckos too large to swallow whole may be treated as a renewable resource, with the 'harvesting' of their tails.

The **Black-headed** or **Freckled Monitor** (*Varanus tristis*) is one Australia's most widespread species, ranging from Perth to Cape York. This slender, long-necked monitor grows to 75 centimetres. It is typically patterned with dark-centred circles over its back and little or no markings over most of the tail. There is a trend for heads and necks to be dark and patternless in the north and west, and spotted in the east. Based on these differences, and on genetic evidence, two subspecies are recognised; the Black-headed Monitor (*V. tristis tristis*) and Freckled Monitor (*V. tristis orientalis*). Yet in some areas they can be difficult to tell apart so colour is a poor means to distinguish

them. The species occurs almost anywhere trees with hollows occur, and it also shelters in rock crevices, under culverts and in human dwellings. They take nestling chicks from hollows and forage widely on the ground for terrestrial lizards.

Spotted Tree Monitors (*Varanus scalaris*) occur in timbered areas across northern Australia, in habitats as diverse as savannah woodlands in the Western Australian Kimberley region to tropical rainforests in Queensland. It is highly likely that this extremely variable monitor, growing to about 60 centimetres, consists of several species. As a group, they are relatively short-limbed and deep-headed, typically with numerous dark-centred spots over their backs. They live in hollow limbs and are extremely wary, often spending many hours resting with just the head exposed.

165

▶ Freckled Monitor
(*Varanus tristis orientalis*).
Wallumbilla, Qld.

▲ Spotted Tree Monitor (*Varanus scalaris*). Badu Island, Qld.

▼ Spotted Tree Monitor. Burdekin, Qld.

Out on a limb: canopy monitors

▲ Emerald Tree Monitor (*Varanus prasinus*). Moa Island, Qld.

Tree-climbing monitors generally confine their activities to trunks and larger branches. They are most comfortable on a relatively broad, rough surface that their claws can cling to. Not so for the canopy monitors. Thanks to their light build, extremely slender bodies and limbs, soft semi-adhesive pads under their feet and very long, thin, prehensile tails these monitors can explore new horizons. They can venture beyond the realm of virtually all other lizards and exploit the thin vines, slender outer branches and foliage.

The two Australian species are restricted to tropical forests in far northern Cape York and the northern Torres Strait islands. They are swift, alert and extremely agile lizards that can move quickly and discreetly through thick vegetation. They also forage on the ground and on tree trunks, and if approached they always keep the trunk between themselves and the observer. When inactive they occupy hollows.

Canopy monitors are opportunistic predators of what ever they can catch and swallow, but their main prey appears to be insects. Prey is immobilised and manipulated in a distinctive manner, being slammed against the substrate and raked with the claws.

The 75 centimetre **Emerald Tree Monitor** (*Varanus prasinus*) is widespread in New Guinea, extending south in Torres Strait to Moa Island. Two sightings 18 months apart from precisely the same vantage point on Moa Peak suggest that individuals reside within home ranges for extended periods.

167

▲ Emerald Tree Monitor (*Varanus prasinus*). Moa Island, Qld.

They are reported to lay their eggs in arboreal termite nests. Considering the diversity of Australian lizard fauna, it is surprising that so few species feature green colouration, particularly in view of the number of green species occupying a variety of available habitats overseas. This monitor is exceptional in being truly bright green.

The **Black Canopy Monitor** (*Varanus keithhorni*) is confined to the Iron and McIlwraith Ranges in far northern Cape York. These areas contain the most significant northern rainforests, which probably comprise core habitat, but the monitors extend beyond these into adjacent eucalypt-dominated open forest. They forage at all levels and use their claws to dig into rotting wood in search of insects.

▲ Black Canopy Monitor (*Varanus keithhorni*). Iron Range, Qld.

Acrobats of the escarpments: rock monitors

▲ Black-palmed Rock Monitor (*Varanus glebopalma*). Kununurra, WA.

Tropical northern and north-western outcrops and escarpments are home to three species of extremely shy rock monitors, among the most superbly structured and agile animals in Australia. They have acute vision thanks to their large eyes set beneath angular brows. Their necks, bodies and limbs are long and slender, and their tails are extremely long and whip-like. These lightning-swift monitors move with ease over vertical surfaces, and sometimes even on rock ceilings beneath overhangs. Undoubtedly these acrobatic predators are formidable adversaries for any smaller creatures inhabiting northern rock faces.

With a total length of 1 metre, the **Black-palmed Rock Monitor** (*Varanus glebopalma*) is the largest rock-inhabiting species. It is also the most widespread, occupying heavily eroded outcrops featuring numerous horizontal crevices and boulder accumulations between the Western Australian Kimberley region and north-western Queensland. This impressive lizard is reddish brown with a black tail tipped with cream to yellow, and a network of dark lines on the neck and throat. Black-palmed Monitors are visually-cued ambush predators that stake out ledges or boulders with a clear field of view. From these vantage points they launch rapid attacks including spectacular leaps on vertebrate and invertebrate prey which are taken back to a crevice to be beaten against the rock and swallowed. It appears their diet shifts according to seasonal availability. During the dry season they take mainly

grasshoppers, moving to lizards at the onset of the wet season, then including frogs as the wet season continues.

Glauert's Monitor (*Varanus glauerti*) occupies gorges and escarpments featuring numerous vertical fissures in far northern Western Australia. When occupying horizontal crevices it prefers those on the tops of outcrops. A separate Arnhem Land population, generally regarded as a similar undescribed species, is strongly arboreal. The 80 centimetre Glauert's Monitor has bands of large grey spots across its back and the tail is strikingly ringed with black and cream. In addition to grasshoppers, prey items feature a large proportion of spiders, cockroaches and lizards, including geckos. The monitor apparently captures much of its food by searching hiding places.

▲ Glauert's Monitor (*Varanus glauerti*). Mitchell Plateau, WA.

▼ Pilbara Rock Monitor (*Varanus pilbarensis*). Python Pool, WA.

Outcrops in the rugged Pilbara region of Western Australia are home to the smallest of the group, the 45 centimetre **Pilbara Rock Monitor** (*Varanus pilbarensis*). This swift and slender species is reddish brown with a black and cream ringed tail and a network of dark lines on the throat. Little is known of its natural history.

Giants in a strange land: large monitors

▲ Perentie (*Varanus giganteus*). North West Cape, WA.

Who can fail to be impressed by the swaggering gait of a huge monitor? They are the largest native terrestrial predators in the land, having outgrown all potential threats – except each other, humans and the introduced Cane Toad. With size comes a degree of complacency bordering on arrogance. Large foraging monitors remain alert and wary, but without the constant prospect of sudden oblivion from the jab of a beak or the swipe of toothed jaws they can safely devote single-minded attention to hunting and scavenging. Using their long flickering tongues they investigate burrows, follow scent trails, and in some cases climb trees to examine hollow limbs. Of course, for juveniles the hazards are many and varied, so hatchlings of our largest monitor species are extremely furtive. The young of even the most common and familiar monitors are rarely seen.

The **Perentie** (*Varanus giganteus*) is the largest Australian lizard, growing to over 2 metres and tipping the scales at a hefty 15 kilograms or more. This unmistakable lizard has distinctive bands of large pale circles across its back, an angular brow, and a long slender neck with a prominent network of dark lines over the throat. Perenties live in the deserts between central Queensland and the west coast, and on islands off the west coast. They are often associated with hilly landscapes featuring gorges and rock outcrops but can also be found in vast tracts of open, featureless terrain. Where natural cavities such as caves and deep fissures are not available they dig extensive burrows about 1 metre deep and up to 8 metres long. They forage widely over home ranges encompassing many hectares, consuming any vertebrates they can catch or carrion they encounter. It

▲ Lace Monitors (*Varanus varius*). Male-to-male combat is a 'push and shove' test of strength. Woodgate NP, Qld.

▼ Lace Monitor juveniles in a termite nest. The brightly coloured young are rarely seen. Sydney, NSW.

is likely they have benefited from introduced rabbits which offer both food and shelter sites. A mummified Perentie resides in the Queensland Museum, having perished after trying to swallow an echidna.

Campers and picnickers in timbered areas of eastern Australia will be familiar with the **Lace Monitor** (*Varanus varius*). For this adaptable lizard, camp grounds offer rich pickings in the form of table scraps. Farmers regard them as nuisances for stealing eggs and chickens. Lace Monitors ('Lacies') grow to nearly 2 metres, with the largest and heaviest individuals occurring at the southern end of the range in Victoria. Although they forage widely on the ground, Lace Monitors take to trees when pursued, spiralling around the trunk to remain hidden from the observer. They also spend a considerable amount of time clinging motionless to tree trunks. Males engage in combat during spring, rearing vertically to grasp each other chest to chest in tests of strength that may leave both protagonists scarred. Eggs are laid in termite mounds which the insects entomb when they seal the damage to their nests. In an extraordinary example of parental care, mothers are said to return at the time of hatching to free the young.

Yellow-spotted Monitors (*Varanus panoptes*) are the stocky heavyweights that patrol much of western and northern Australia. There are two subspecies, with *V. panoptes panoptes* across the north and *V. panoptes rubidus* on the west coast and interior of Western Australia. Yellow-spotted Monitors grow to about 1.5 metres. They typically have alternating bands of black and yellow spots across their backs, black spots on their bellies and black blotches on their throats. They are terrestrial, sheltering in large burrows and hollow logs. In some northern areas they are often associated with floodplains, but their broad habitats range from open forest to beaches. They feed on what ever they can catch or scavenge, with recorded prey including other monitors, aquatic snakes, death adders and turtles' eggs. According to seasonal prey availability, they have been observed to forage over 6 kilometres in a day.

▼ Yellow-spotted Monitor (*Varanus panoptes rubidus*) swallowing a dragon. Twin Peaks, WA.

▲ Yellow-spotted Monitor (*Varanus panoptes panoptes*). Aramac, Qld.

▲ Gould's or Sand Goanna (*Varanus gouldii*). Sturt NP, NSW. G. Swan

The **Gould's** or **Sand Goanna** (*Varanus gouldii*) is Australia's most widespread monitor, occurring in large areas of all mainland states, but continues to be confused with the more robust Yellow-spotted Monitor. They grow to similar lengths in the north, but in the south the Sand Goanna is smaller, reaching just over 1 metre. It lacks the black spots on the belly, and has a grey V-shape on the throat. Where the two species occur together, the Sand Goanna generally occupies softer, sandy soils. When pursued, Sand Goannas retreat to burrows, usually less than 1 metre deep and 2–5 metres long, with a couple of branches ending in terminal chambers often within 30 centimetres of the surface. A 'pop-hole' is constructed to the surface, meaning they can rapidly 'burst from the ground' and escape if their burrow is being excavated. When foraging they often stand on their hind limbs to survey their surroundings.

▲ Gould's or Sand Goanna. Coolgardie area, WA.

Best of both worlds: water monitors

▲ Merten's Water Monitors (*Varanus mertensi*) are fast disappearing from tropical waterways as Cane Toads spread across the Top End. Kununurra, WA.

Tropical northern waterways lined with overhanging trees and pandanus or mangroves are home to four species of semi-aquatic water monitors. These are capable climbers that spend a great deal of time clinging to trunks and branches, but are seldom found far from water. They forage along creek banks and swamp edges and are excellent swimmers that include aquatic prey in their diets. If approached from the water they tend to retreat higher into trees, but if threatened from land they do not hesitate to leap into the water and can remain submerged for extended periods.

Adaptations to an aquatic lifestyle vary between species. They include a laterally-flattened tail for swimming, and in some cases the placement of the nostrils on the top of the snout. This allows the animal to remain underwater and breathe with only a small portion of its snout breaking the surface.

Merten's Water Monitor (*Varanus mertensi*) is drab dark grey with fine pale spots and grows to a total length of just over 1 metre. It is Australia's most aquatic goanna, with a very strongly laterally-compressed tail and valvular nostrils that can be closed underwater. It has even been recorded to forage underwater, walking across the bottom investigating nooks and crannies in the same way it does on land. It is also known to excavate the submerged nests of the Northern Long-necked Turtle (*Chelodina rugosa*). Freshwater crabs are an important component of its broad diet. It extends across northern Australia between Cape York and the Kimberley region.

The 70 centimetre **Mitchell's Water Monitor** (*Varanus mitchelli*) shares the riverine habitats with its larger relative between far western Queensland and the

175

◄ Merten's Water Monitor (*Varanus mertensi*). Kununurra, WA.

▼ Rusty Monitor (*Varanus semiremex*). Mid-eastern Qld.

Kimberley region of Western Australia. It is an extremely wary species that shelters in tree hollows, pandanus roots, rock crevices and behind loose bark. Both it and Merten's Water Monitor frequently make use of bridges and culverts, where they can often be found sleeping at night. During the wet season it extends along small creeks and inundated paperbark forests, contracting back to permanent billabongs and large rivers during the dry season.

The **Rusty Monitor** (*Varanus semiremex*) mainly occupies mangrove forests along brackish estuaries, extending inland 50-100 km beside freshwater streams and paperbark swamps. It ranges from mid-eastern Queensland to western Cape York. This 60 centimetre lizard has dark, smoothly polished scales on top of its head and usually a rusty yellow flush on the chest and throat. Its aquatic modifications are not as pronounced as in other water monitors, with the basal third of its tail round in cross-section and only the rear two-thirds being laterally compressed. Lizards living in mangrove areas are able to rely heavily on marine crabs, thanks to glands that enable them to eliminate excess salt through the nostrils.

▲ Mitchell's Water Monitor (*Varanus mitchelli*). Kununurra, WA.

Suggested reading

Aplin, K.P. & Smith, L. A., 2001. 'Checklist of the frogs and reptiles of Western Australia'. *Records of the Western Australian Museum.* Supplement No. 63: 51–74.

Bennett, R., 1997. *Reptiles and Frogs of the Australian Capital Territory.* National Parks Association of the ACT Inc, Canberra.

Bush, B., Maryan, B., Browne-Cooper, R. & Robinson, D., 1995. *A Guide to the Reptiles and Frogs of the Perth Region.* University of Western Australia Press, Perth.

Bush, B., Maryan, B., Browne-Cooper, R. & Robinson, D. 2007. *Reptiles and Frogs in the Bush: Southwestern Australia.* University of Western Australia Press, Perth.

Cogger, H.G., 2000. *Reptiles and Amphibians of Australia.* Reed New Holland, Sydney.

Ehmann, H., 1992. *Encyclopedia of Australian Animals. Reptiles.* Collins Angus & Robertson, Sydney.

Green, K. & Osborne, W., 1994. *Wildlife of the Australian Snow Country.* Reed Books, Chatswood.

Greer, A. E., 1989. *The Biology and Evolution of Australian Lizards.* Surrey Beatty & Sons, Chipping Norton, NSW.

Griffiths K., 2006. *Frogs and Reptiles of the Sydney Region.* Reed New Holland, Sydney.

Horner, P., 1991. *Skinks of the Northern Territory.* Northern Territory Museum of Arts and Sciences, Darwin.

Hoskin, C.J., Couper, P.J. & Schneider, C.J., 2003. 'A new species of *Phyllurus* (Lacertilia: Gekkonidae) with a revised phylogeny and key for the Australian leaf-tailed geckos.' *Australian Journal of Zoology* 51: 153-64

Houston, T., 1998. *Dragon Lizards and Goannas of South Australia.* Revised by M. Hutchinson. South Australian Museum

Hutchinson, M., Swain, R. & Driessen, M., 2001. *Snakes and Lizards of Tasmania.* University of Tasmania, Hobart.

Jenkins, R. & Bartell, R., 1980. *A Field Guide to Reptiles of the Australian High Country.* Inkata Press, Melbourne.

Pianka, E., King, D.R. & King, R.A., 2004. *Varanoid Lizards of the World.* Indiana University Press, USA.

Storr, G.M., Smith, L.A. & Johnstone, R.E., 1983. *Lizards of Western Australia II: Dragons and Monitors.* Western Australian Museum. Perth.

Storr, G.M., Smith, L.A. & Johnstone, R.E., 1990. *Lizards of Western Australia III: Geckos and Pygopods.* Western Australian Museum. Perth.

Storr, G.M., Smith, L.A. & Johnstone, R.E., 1999. *Lizards of Western Australia I: Skinks.* Western Australian Museum. Perth.

Swan G., Shea, G. & Sadlier, R., 2004. *A Field Guide to Reptiles of New South Wales.* Reed New Holland, Chatswood.

Swan, M. & Watherow, S., 2005. *Snakes, Lizards and Frogs of the Victorian Mallee.* CSIRO Publishing.

Wilson, S., 2005. *A Field Guide to Reptiles of Queensland.* Reed New Holland, Chatswood.

Wilson, S., 2008. *Reptiles of the Southern Brigalow Belt.* Rev. edn. World Wildlife Fund, Australia.

Wilson, S. & Knowles, D., 1988. *Australia's Reptiles: A Photographic Reference to the Terrestrial Reptiles of Australia.* Collins, Sydney.

Wilson, S. & Swan, G., 2008. *A Complete Guide to Reptiles of Australia.* Rev. edn. Reed New Holland, Chatswood.

Index

Acknowledgements

Steve Wilson thanks his wife, Marilyn Parker, for her continued support. Living with an obsessive compulsive herpetologist cannot be easy. His parents, Joy and Ken urged him from an early age to follow his dreams. He thanks Bob Ashdown, Rod Hobson, Dave Knowles and Mike Swan for stimulating time spent together in the bush, and thanks Kieran Aland, Mark Fitzgerald, Tony and Kate Hiller and Rex Neindorf for access to photographic subjects.

Gerry Swan thanks his family for their patience and support, particularly his wife Marlene who is much more clever at spotting those elusive nondescript brown skinks in the field. Thanks also to Steve Sass, Mats Olsson, Brett Aitchison and Ross Sadlier for good times in the bush.

We are grateful to NACAP for fostering scientific research, including the documentation and photography of reptiles during pipeline construction, and we thank Terry Annable, Brian Bush, Steve Donnellan, Paul Horner, Dave Knowles, Brad Maryan and Geoff Swan for the use of their photographs.